Proceedings in Life Sciences

Fungal Viruses

XIIth International Congress
of Microbiology, Mycology Section,
Munich, 3-8 September, 1978

Proceedings of the Symposium on Fungal Viruses
Including Abstracts of Papers of the Symposium on
Extrachromosomal Vectors in Fungi and
Abstracts of Posters on Fungal Viruses

Edited by
H. P. Molitoris M. Hollings H. A. Wood

With 78 Figures

Springer-Verlag
Berlin Heidelberg GmbH 1979

Professor Dr. H. P. Molitoris
Institut für Botanik II
Arbeitsgruppe Pilzphysiologie
Universität Regensburg
Universitätsstraße 31
8400 Regensburg/FRG

Dr. M. Hollings
Head Virology Department
Glasshouse Crops Research Institute
Worthing Road, Littlehampton
West Sussex BN16 3PU/GB

Dr. H. A. Wood
Boyce Thompson Institute for Plant Research
at Cornell University
Tower Road
Ithaca, NY 14853/USA

ISBN 978-3-642-67375-7   ISBN 978-3-642-67373-3 (eBook)
DOI 10.1007/978-3-642-67373-3

Library of Congress Cataloging in Publication Data. International Congress for Microbiology, 12th, Munich, 1978. Mycology Section. Fungal viruses. (Proceedings in life sciences) Bibliography: p. Includes index. 1. Fungal viruses–Congresses. 2. Fungal genetics–Congresses. I. Molitoris, H. Peter. II. Hollings, M., 1923–. III. Wood, Harry Alan, 1941–. IV. Title. QR343.I57. 1978. 576'.6483. 79-9437.

© Springer-Verlag Berlin Heidelberg 1979

Originally published by Springer-Verlag Berlin Heidelberg New York 1979

Softcover reprint of the hardcover 1st edition 1979

2131/3130-543210

Preface

This book records the contributions presented at the XIIth International Congress of Microbiology, Mycology Section, held in Munich on 3–8 September 1978. All the papers given at the Symposium (no. 33) on Fungal Viruses, and at the Round Table Discussion (RTD 1) are reported in full, and the paper on fungi as vectors of plant viruses by R.N. Campbell (who was unfortunately unable to attend the Congress) has also been included (Part A).

Much of the current work with viruses in fungi involves genetic studies and virus-host gene interactions; for this reason, the Symposium (no. 32) on Extrachromosomal Vectors in fungi has also been reported, in the form of abstracts from all those contributors who gave permission for this (Part B).

Authors' abstracts of posters relating to fungal viruses have similarly been given (Part C).

Fungal viruses, or mycoviruses, can be defined as viruses that replicate in fungi, and since their discovery in 1962, considerable progress has been made towards an understanding of their biological and particularly their physico-chemical properties. Present knowledge suggests that mycoviruses are usually latent in nature, and their biological manifestations can often be more readily studied as cytoplasmically inherited determinants than as viruses. An attempt has therefore been made in this book to bring together the most recent advances, not only in mycovirus research, but also in extrachromosomal determinants in fungi.

Mycoviruses have been found in fungi from all the major taxonomic groups. Most of these viruses have been described only by electron microscopy, and they are therefore usually referred to as virus-like particles (VLP). During the last few years, however, purification procedures have been applied to more of these viruses, and many have now been characterized *in vitro* sufficiently to indicate their typical virus nature: these are generally referred to as mycoviruses. The general lack of transmission techniques has largely precluded the demonstration of infectivity of mycoviruses.

Although mycoviruses have been studied in only a limited number of laboratories around the world, it has become evident that these viruses are very common; the great majority of those examined have been shown to contain a double-stranded ribonucleic acid (dsRNA) genome, and they thus represent the largest recognized group of dsRNA viruses. Taxonomically and biologically, however, they appear to differ considerably from the dsRNA viruses of the family Reoviridae. Whereas viruses in the Reoviridae are considered to be disease agents, there is very little adequately controlled evidence as to whether most mycoviruses are truly latent, or whether they induce more subtle cytopathic effects in their hosts. Mycoviruses appear to be intimately associated with their hosts and in some instances they control or modify certain biological attributes of the host cell; as such, they are an important part of mycology.

In the early days of mycovirus research, several topics were investigated because of their immediate practical application in the fields of pharmacology and medicine. The possible influence of mycoviruses on antibiotic production in *Penicillium* species and other industrial fungi, and on the production of mycotoxins by spoilage fungi of food products, received much attention. In both these important aspects, however, mycoviruses appeared to exert no significant effect. The potential clinical or veterinary uses of the dsRNA of mycoviruses as an interferon-inducing agent were explored, but the toxic side-effects soon discouraged further investigation. More recently, the possibilities have been examined of using mycoviruses that infect plant-pathogenic fungi as a means of biological control of these fungi. So far, there has been no consistent evidence that the observed hypovirulence in these fungi is actually caused by mycovirus infection, and this problem is further complicated by the difficulties in experimental transmission of the viruses concerned.

This lack of suitable techniques for experimental transmission remains the most serious limitation to mycovirus research today, and it is therefore all the more remarkable that so much information on the replicative strategies of different mycoviruses within the host cell has been elucidated, and that the complexities of the killer systems in *Saccharomyces* and *Helminthosporium* species have been at least partly unravelled. The killer systems have provided a unique experimental tool for studying the interactions of determinants carried in the mycovirus genome and those carried in the fungal chromosomes and cytoplasm.

Besides the mycoviruses, which infect fungi, there are a number of plant viruses that are transmitted by Chytrid fungi, although there is no evidence that any of these plant viruses can multiply in its fungal vector. The subject of fungi as vectors of plant viruses is included in the scope of this book.

The full biological significance of mycoviruses to their hosts can not yet be evaluated, but their very prevalence and efficient perpetuation ensure that they cannot be ignored by any who study fungi. This book represents an overview of the present state of our knowledge.

Thanks are due to the many people whose willing help and cooperation have made possible the production of this book. The generous financial assistance of the Deutsche Forschungsgemeinschaft, Bonn, is acknowledged; this not only made it possible for a number of overseas speakers to attend the symposium in Munich, but also enabled H.P.M. to attend the organisational meeting during the 2nd International Mycological Congress in Tampa, Florida.

Finally, it is a pleasure to thank Miss Eva Grajf for her careful help in typing and proofreading and the editorial staff of Springer-Verlag for their kind cooperation and expert assistance.

October, 1979

H.P. MOLITORIS
M. HOLLINGS
H.A. WOOD

Contents

Methods

Taxonomy

Part B
Symposium on Extrachromosomal Vectors in Fungi — Abstracts
Chairman: E.A. BEVAN
Convenor: K. ESSER

Part C
Posters on Fungal Viruses – Abstracts

Contributors

You will find the addresses at the beginning of the respective contribution

Part A
Symposium and Round Table Discussion on Fungal Viruses

Chairmen: M. HOLLINGS and H. A. WOOD
Convenor: H. P. MOLITORIS

Evolution

Coevolution of Fungi and Their Viruses

P. A. LEMKE

Mellon Institute, Carnegie-Mellon University, Pittsburgh, PA 15213/USA

1 Introduction

Evolution is a safe subject for discussion, as one can afford to speculate without cause for contradiction. A discussion on the coevolution of fungi and their viruses will be especially speculative, as it is based on the biological attributes of present-day fungi and on the properties of rather few viruses, those double-stranded RNA (dsRNA)-containing mycoviruses that have been physicochemically well characterized.

2 Virus-Host Coevolution

The physicochemical properties of dsRNA mycoviruses relative to dsRNA viruses of nonfungal hosts have been compared previously by Wood (1973). The biological properties of the dsRNA mycoviruses have been discussed extensively by others (Bozarth, 1972; Lemke and Nash, 1974; Lemke, 1976a; Hollings, 1978; Saksena and Lemke, 1978) and indicate that most if not all mycoviruses are heritable viruses — endogenous to the host cell not only during replication but during transmission. Their transmission occurs either during cell division (serial transmission) or following cell fusion (lateral transmission). Serial transmission is accomplished effectively in those fungi producing spores such as conidiospores (mitotic spores). Indeed many fungi, sexual as well as a-sexual fungi, produce such spores in great profusion. Other fungi produce spores of different types such as basidiospores (meiotic spores) or arthrospores (mitotic spores produced by fragmentation of vegetative cells). Regardless of spore type, all fungal spores, if they incorporate and retain virus particles in situ, are potential agents for serial transmission. Lateral transmission accompanies plasmogamy between genetically compatible fungi either during mating of sexually compatible strains or during fusion between vegetatively compatible strains regardless of their potential for sexual reproduction. Vegetative incompatibility, either within the species or between species, promotes inbreeding, and such inbreeding would delimit lateral transmission of mycoviruses. It is indeed interesting to speculate in this context that mycoviruses, as heritable agents, could well have afforded the selective pressure for inbreeding and perhaps even loss of sexual competence among fungi. Inbreeding mechanisms are common in fungi and have been discussed previously by the author (Lemke, 1973) and in textbooks devoted to fungal genetics (Esser and Kuenen, 1967; Burnett, 1975). There are thus far no adequate models to explain the selective advantage for inbreeding and apomixis in fungi.

The endogenous and noninfectious nature of mycoviruses characterized to date presupposes a compromise — an infection that is both latent and persistent. Latency benefits the host for survival, and persistence benefits the virus in the absence of infectivity. Clearly this compromise, while not typical of virus-host relationships, might well have evolved in response to an opportunity for more efficient viral transmission, lateral as well as serial, when integrity of a host system is maintained. Fungal systems may well have afforded this opportunity by their characteristic cellular organization, their potential for cytoplasmic exchange between vegetative cells, and their efficient reproduction through spores.

Host-mediated transmission, however, might lead to considerable loss of integrity in organization of a virus, and this indeed seems evident from physicochemical details on mycoviruses (Wood, 1973; Lemke, 1976a). The dsRNA mycoviruses, unlike other dsRNA-containing viruses, are frequently multicomponent (i.e., genome segments are distributed among a population of particles rather than in a single particle). Supernumerary molecular weight forms of dsRNA, including dsRNA molecules nonessential for production of virus particles, exist for some fungal viruses (Vodkin et al., 1974; Adler et al., 1976; Koltin, 1977; Lemke, 1977). Data on sedimentation of viral components indicate that large numbers of particles may be empty, and in one fungal system, *Aspergillus flavus* (Wood et al., 1974), all particles assembled are devoid of nucleic acid. These latter particles truly deserve to be called "viruslike particles."

2.1 Virus Concept

Since the fungal viruses characterized to date do not satisfy well all criteria expected of a virus, many investigators have adopted the term "viruslike particles" or "VLP" for these particles. Use of VLP, however, simply avoids the issue as to the nature of such particles. The term "defective" viruses has been considered, but this term implies degeneracy and is unfortunate, since the viruses of fungi, although relatively simple in structure, may be highly evolved and well-adapted to their host. A case for this will be presented below. In this paper, the term "heritable" viruses is preferred as rather descriptive of fungal viruses, which have thus far proven to be infectious only through heredity. Ultimately, fungal viruses may be found that are routinely infectious, and such viruses would of course not fit this description.

2.2 Virus Evolution as Evidenced by Reduction

Evolution from more complex progenitors to simpler forms, through reduction or loss in structure and function, has occurred repeatedly in the evolution of life forms, and for higher organisms this is often documented in the fossil record. Evidence for evolution by reduction of fungal viruses is indirect but can be gleaned from the comparative physicochemical properties of several fungal virus systems. For example, two very closely related fungi, *Penicillium cyaneofulvum* and *Penicillium chrysogenum*, each contains serologically identical viruses for which the capsid proteins and amino acid composition are identical (Buck and Girvan, 1977). These two viruses, however, differ

in genome complexity, as the virus of *Penicillium cyaneofulvum* contains an extra segment of dsRNA. Extra and apparently superfluous dsRNA segments are found also in
the viruses of *Saccharomyces cerevisiae* (Vodkin et al., 1974; Adler et al., 1976) and
Ustilago maydis (Koltin, 1977) isolated from specific host strains. Indeed, determinants
for phenotypes other than virus structure and replication such as the toxin and immunity determinants in the killer systems of these fungi reside in such extra segments. In the
light of the theory for evolution by reduction, these killer systems might well be regarded as dynamic systems in which such an evolution is still in progress.

It has already been suggested in this paper that an endogenous or heritable virus, in
addition to loss of infectivity, could attain considerable reduction and modification in
structure. The *Penicillium stoloniferum* slow virus, for example, which has been well
characterized both for structure and replication (Buck and Kempson-Jones, 1973, 1974;
Buck, 1975; and see the chapter by Buck in this volume) is surely to be considered a
simple virus, although there is no direct evidence to prove that its simplicity evolved
by reduction. Regardless, this virus has only two genome segments – one sufficiently
large to encode for the single major capsid polypeptide and one sufficiently large to
encode for the viral specific replicase. There are no excesses. This heritable virus is a
model for adaptation for endogenous replication; it retains a minimal genome for
maintenance of what might still qualify as a virus (i.e., a nucleoprotein particle able to
replicate in a host cell). Further reduction might lead to a lone genome segment able
to encode for its replicase. This has not been identified in fungal cells, but such a determinant, admittedly nonviral in organization, is conceivable as the ultimate product
of viral evolution via reduction.

With reference to fungal viruses, three points deserve to be mentioned here. First,
the spectrum of dsRNA mycoviruses indicates considerable heterogeneity for these
viruses both in serology and genome segmentation (Wood, 1973; Lemke, 1976a; and
see chapter by Bozarth in this volume). Thus, if reduction has occurred in the evolution of fungal viruses, it has apparently occurred repeatedly from several lines of viral
progenitors. Secondly, in view of the prevalence of dsRNA-containing viruses in fungi,
it would appear that the dsRNA genome may be preferentially suited for reduction
leading to endogenous and persistent infection characteristic of a heritable virus. Indeed, models for replication of dsRNA fungal viruses indicate replicative cycles that are
abbreviated when compared with the replication of dsRNA viruses from nonfungal
hosts. The replication of fungal viruses, as heritable and persistent viruses, may well
have been modified to keep pace with but not to exceed the rate of host cell division.

2.3 Host Evolution as Evidenced by Nuclear Gene Mutations

Indicative evidence that fungi and their viruses have coevolved and are coadapted to a
considerable degree rests on the latent or symptomless relationship of most fungal viruses with their hosts. Specific evidence for coadaptation has been obtained from genetic studies with the yeast *Saccharomyces cerevisiae* (Wickner, 1974, 1978; Wickner and
Leibowitz, 1976) and the mold *Penicillium chrysogenum* (Lemke et al., 1976). In the
yeast there are more than 20 nuclear genes required for maintenance of the dsRNA
segment encoding for determination of killer phenotype. This extra dsRNA segment,

specifically, and not the virus of yeast is maintained by the wild-type or dominant alleles of these nuclear genes; mutation leads to loss of this extra dsRNA. This latter point is significant in view of the already discussed potential for reduction in the evolution of fungal viruses. The studies with *Penicillium chrysogenum* indicate that the dominant allele of at least one nuclear gene is required for maintenance of the host when virus titer is high (Lemke et al., 1976). In the wild-type *Penicillium chrysogenum* the association of virus and host is symptomless as in most fungal viruses. Only a mutant strain of *Penicillium chrysogenum,* grown on a lactose-containing medium to enrich for virus titer, showed tissue degeneration or conditional lysis. The general absence of lysis or of disease symptoms in fungi containing viruses may be related not only to evolutionary reduction of these viruses but to evolution of host genes to accommodate endogenous viral infection. The absence of such evolution on the part of the host could lead to a pathogenic response to viruses, as is observed in the case of the mushroom virus disease of *Agaricus bisporus* (Hollings, 1962; Dieleman-van Zaayen, 1972; Lemke, 1976b).

2.4 Gene Frequencies and Heritable Viruses

The only mathematical models for evolution (change in gene frequency) of heritable viruses are based on analysis of certain viruses, principally the sigma virus, in populations of *Drosophila* (L'Héritier, 1970). Basically, three conclusions were drawn from this study, and these are mentioned here only because of their possible relevance to the study of fungal viruses. First, if inheritance of a virus is strictly maternal but incomplete (i.e., the virus is never transmitted through the sperm and is occasionally absent from the egg), then a totally noninfectious virus in order to be retained in a breeding population must confer some selective advantage upon the host. Secondly, if transmission is bidirectional and efficient (i.e., either parent may transmit the virus or, as in the case of many fungi, where cytoplasmic exchange is not limited to gametes and is possible through vegetative cell fusion), then a heritable virus may persist and indeed spread even without a selective advantage. By extrapolation, this second set of circumstances may well reflect what is common for mycoviruses, viral latency yet persistence without the aid of natural selection. Thirdly, heritable viruses, if they are detrimental or pathogenic to the host, and indeed noninfectious, should be lost readily from a breeding population, especially if the viral determinants are extrachromosomal and not linked to essential genetic elements. Of special relevance in this context, and seemingly contradictory, is the mushroom virus disease of *Agaricus bisporus,* a disease that is correlated with the presence of double-stranded RNA and with several morphological types of virus (Hollings, 1962; Dieleman-van Zaayen, 1972; Saksena, 1975; Marino et al., 1976; Lemke, 1976b; Del Vecchio et al., 1977, 1978). This disease involves heritable viruses that are pathogenic yet persistent and widespread among cultivated strains of *Agaricus bisporus.* It might be suggested either that heritable determinants for this disease are linked to essential genetic factors and are thereby retained (i.e., perhaps integrated DNA copies of these viruses exist in *Agaricus bisporus*) or that mushroom viruses, like other fungal viruses, are basically latent but are allowed to reach unnatural or pathogenic levels during the process of mushroom cultivation. Further study of mushroom virus disease should focus upon such alternatives. The infectivity of mushroom viruses as free particles, considered to be negligible to absent on the basis of early work, should also now be reinvestigated in the light of more sensitive assays for virus detection (see chapters by Del Vecchio et al. and Lister in this volume). If mushroom viruses are indeed infectious, then the contradiction alluded to above does not exist.

3 Concluding Remarks

This essay depicts the known fungal viruses as heritable elements that have evolved by reduction in structure and function, as well as in potential for infectivity, from an extended series of more complex viruses. It is doubtful that details in this evolution can ever be reconstructed, but the serological and physicochemical data available to date indicate that this evolution occurred repeatedly among unrelated viruses and selectively among viruses with double-stranded RNA genomes. These events, although not retraceable, have led investigators of fungal viruses to reevaluate those criteria normally attributed to a virus. The fungal viruses, with perhaps one or two known exceptions, do not represent pathogenic or infectious agents; they are in concert with their host. Such associations could only have arisen through coevolution of fungi and their viruses, and each association should be studied seriously if we are to develop a classification for fungal viruses that is both meaningful and natural.

Summary

Fungal viruses are principally dsRNA-containing and endogenous to their host. Most, if not all, are transmitted laterally by cytoplasmic exchange or serially through spores or other propagules. Since they are infectious (lateral transmission) by cytoplasmic heredity, their spread is defined by incompatibility genes, especially genes for vegetative incompatibility. These viruses, despite similarity in appearance and simple structure, are heterogeneous for genome segmentation and serological relationships. Such viruses may be simple by reduction and represent residual forms of more complex progenitors. Their endogenous nature implies a close interaction with host genes for maintenance and replication, as well as lateral transmission. Fungal viruses offer model systems for study of nucleo-cytoplasmic interactions in eukaryotic cells. Any evidence for informational DNA copies of dsRNA segments would be most exciting, since this implies reverse transcriptase activity and the possibility of vector relationships involving recombinant DNA.

References

Adler J, Wood HA, Bozarth RF (1976) Virus-like particles in killer, neutral, and sensitive strains of *Saccharomyces cerevisiae*. J Virol 17:472–476

Bozarth RF (1972) Mycoviruses: a new dimension in microbiology. In: Environmental health perspectives. US Dept Health, Education and Welfare, Washington, DC, pp 23–29

Buck KW (1975) Replication of double-stranded RNA in particles of *Penicillium stoloniferum* virus S. Nucl Acids Res 2:1889–1902

Buck KW, Girvan RF (1977) Comparison of the biophysical and biochemical properties of *Penicillium cyaneo-fulvum* virus and *Penicillium chrysogenum* virus. J Gen Virol 34:145–154

Buck KW, Kempson-Jones GF (1973) Biophysical properties of *Penicillium stoloniferum* virus S. J Gen Virol 18:223–235

Buck KW, Kempson-Jones GF (1974) Capsid polypeptides of 2 viruses isolated from *Penicillium stoloniferum*. J Gen Virol 22:441–445

Burnett JH (1975) Mycogenetics. Wiley, London, pp 375

Del Vecchio VG, Dixon C, Lemke PA (1977) Immunoelectrophoretic detection of double-stranded ribonucleic acid from *Agaricus bisporus*. Exp Mycol 1:102–106

Del Vecchio VG, Dixon C, Lemke PA (1978) Immune electron microscopy of virus-like particles of *Agaricus bisporus*. Exp Mycol 2:138–144

Dieleman-van Zaayen A (1972) Mushroom virus disease in the Netherlands: symptoms, etiology, electron microscopy, spread and control. Centre for Agric Publ and Document, Wageningen, Netherlands, pp 130

Esser K, Kuenen R (1967) Genetics of fungi. Springer, Berlin, Heidelberg, New York, pp 500

Hollings M (1962) Viruses associated with a die-back disease of cultivated mushroom. Nature (London) 169:692–695

Hollings M (1978) Mycoviruses: viruses that infect fungi. Adv Virus Res 22:1–53

Koltin Y (1977) Virus-like particles in *Ustilago maydis:* mutants with partial genomes. Genetics 86:527–534

Lemke PA (1973) Isolating mechanisms in fungi-prezygotic, postzygotic, and azygotic. Persoonia 7:249–260

Lemke PA (1976a) Viruses of eucaryotic microorganisms. Annu Rev Microbiol 30:105–145

Lemke PA (1976b) Fungal viruses and agriculture. In: Romberger JA (ed) Virology in agriculture. Allanheld Osmun, Montclair, New York, pp 159–175

Lemke PA (1977) Double-stranded RNA viruses among filamentous fungi. In: Schlessinger D (ed) Microbiology–1977. Am Soc Microbiol, Washington DC, pp 568–570

Lemke PA, Nash CH (1974) Fungal viruses. Bacteriol Rev 38:29–56

Lemke PA, Saksena KN, Nash CH (1976) Viruses of industrial fungi. In: Macdonald KD (ed) Genetics of industrial microorganisms. Academic Press, New York, pp 323–355

L'Héritier P (1970) *Drosophila* viruses and their role as evolutionary factors. Evol Biol 4:185–209

Marino R, Saksena KN, Schuler M, Mayfield JE, Lemke PA (1976) Double-stranded ribonucleic acid in *Agaricus bisporus*. Appl Envir Microbiol 31:433–438

Saksena KN (1975) Isolation and large-scale purification of mushroom viruses. Dev Ind Microbiol 16:134–144

Saksena KN, Lemke PA (1978) Viruses in fungi. In: Fraenkel-Conrat H, Wagner RR (eds) Comprehensive virology, vol 12. Plenum, New York, pp 103–143

Vodkin M, Katterman F, Fink GR (1974) Yeast killer mutants with altered double-stranded ribonucleic acid. J Bacteriol 117:681–686

Wickner RB (1974) Chromosomal and nonchromosomal mutations affecting the "killer character" of *Saccharomyces cerevisiae*. Genetics 76:423–432

Wickner RB (1978) Twenty-six chromosomal genes needed to maintain the killer double-stranded RNA plasmid of *Saccharomyces cerevisiae*. Genetics 48:419–425

Wickner RB, Leibowitz MJ (1976) Chromosomal genes essential for replication of double-stranded RNA plasmid of *Saccharomyces cerevisiae:* the killer character of yeast. J Mol Biol 105:427–443

Wood HA (1973) Viruses with double-stranded RNA genomes. J Gen Virol 20:61–85

Wood HA, Bozarth RF, Adler J, Mackenzie DW (1974) Proteinaceous virus-like particles from an isolate of *Aspergillus flavus*. J Virol 13:532–534

Fungi as Vectors and Hosts of Viruses

Fungal Vectors of Plant Viruses

R.N. CAMPBELL

Department of Plant Pathology, University of California, Davis CA 95616/USA

1 Introduction

This paper is concerned with plant viruses (or viruslike agents) that are soil-borne and particularly with those having a fungal vector. These viruses are intimately associated with the fungus, often being harbored within the fungal resting spore. Nevertheless, they do not seem to multiply within the fungus. The emphasis in this paper will be on the fungus—virus relationships that are of biological and epidemiological interest.

2 General Characteristics of Fungal Transmission

2.1 Definitions of Soil-Borne and Soil-Transmitted Viruses

In this paper "viruses" will mean plant viruses and viruslike agents that cause plant diseases. A distinction will be made between "soil-borne" viruses and "soil-transmitted" viruses. The broader term "soil-borne" will denote a virus that is introduced into the soil and that subsequently infects healthy plants. The virus can be introduced into soil by incorporation of infected plant material, by release from the roots of living plants, by experimental addition to the soil as a purified or partially purified preparation, or by a soil-borne vector. All plant viruses enter the soil with infected host tissue. For most, however, this constitutes a "dead-end" as they have no effective means of surviving or being inoculated into succeeding host plants. Tobacco mosaic virus, to be discussed later, is the prime example of a virus adapted to this method. Experimentally it is also possible to add relatively large amounts of virus to the soil in the form of purified or partially purified preparations and to demonstrate infection of susceptible plants. Presumably infection occurs because the virus has saturated the root surface and there is mechanical damage to cells of the root system. This kind of damage is possible in spite of attempts to avoid it if the pots are moved about or even if only the tops of plants are moved resulting in physical abrasion or damage to root cells, or if there is injury from the soil fauna. Growth of the root tip or root hairs in soil does not by itself result in injuries that can serve as infection sites. A comparable type of soil-borne infection occurs when susceptible plants are removed from pots and new plants are transplanted into the same soil. In this case the mechanical transmission to the roots is obvious. While the virus in either case can be called soil-borne, this hardly constitutes evidence that the virus is soil-borne in nature or that the virus should be expected to have a soil-borne vector. The term "soil-transmitted" will be used when there is clear evidence that a soil-infesting vector is responsible for transmission of a virus to the roots of plants. Ironically, the virus that is inoculated into the host roots in most cases is not free in the soil but is within the vector.

Although diseases that are recognized today as soil-borne viral diseases were recognized nearly 100 years ago, this is the 20th anniversary of the current era. In 1958, two papers from the University of California at Davis associated *Xiphinema index* with the transmission of grapevine fanleaf virus (Hewitt et al., 1958) and *Olpidium brassicae* with lettuce big-vein disease (Grogan et al., 1958). Since then the soil-transmitted viruses have been grouped into those that are nematode-transmitted and those that are fungus-transmitted. Recent summaries of nematode transmission are available (Lamberti et al., 1975; Taylor and Robertson, 1977).

2.2 Characteristics of Fungal Vectors

Fungal vectors have been demonstrated for several plant viruses and it is now possible to make some general observations on the nature of these interesting relationships. First, there is as high a degree of specificity as with other types of vectors. Viruses transmitted by fungi are not transmitted by other types of vectors and a given virus is only transmitted by one fungal vector species. Second, these viruses infect and multiply in the host but apparently not in the fungus. Third, the vectors identified to date are in the genera *Olpidium* and *Synchytrium* of the Chytridiales and *Polymyxa* and *Spongospora* of the Plasmodiophorales. Fourth, the virus—vector relationship may be one of two types. I shall amplify the latter two observations before turning to a survey of known or suspected cases of fungal transmission.

The documented fungal vectors share many characteristics. They are obligately parasitic, root-infecting, lower fungi that may be pathogens in their own right or simply parasites. They produce zoospores that are uni- or bi-flagellate in the Chytridiales or Plasmodiophorales, respectively. The zoospores encyst on and infect the host cells. Ultrastructural studies with *Olpidium brassicae* have shown that for the first 36 h the fungus develops within the host protoplast as a thallus bounded only by the thallus ectoplast (Temmink and Campbell, 1968, 1969b; Lesemann and Fuchs, 1970). During this time the fungus grows without obvious damage to the host protoplast. Judging from light microscopic observations of other vectors, their ultrastructural relationship to their host is likely to be similar except that some pathogens cause hypertrophy and hyperplasia or that longer times may pass before the thallus matures. Finally, the fungi produce vegetative sporangia that release additional zoospores and thicker walled resting spores. The latter are essential for survival of the fungus in the absence of a host and usually also for survival of the virus. The cystosori of the Plasmodiophorales will be referred to as resting spores in this paper.

It is hypothesized that two of these features are essential to the vector capability of any fungus: (1) the existence of an ectoplast-limited thallus that permits exchange of virus between host and vector and (2) the obligately parasitic relationship in which the host protoplast continues to function more or less normally, thereby providing a suitable milieu for virus infection. It is difficult to visualize either how a virus could be moved through the cell wall of a fungus that infects by means of a germ tube or hypha or how the virus could successfully infect a protoplast that is rapidly killed by the penetrating germ tube or hypha.

The obligately parasitic nature of the fungal vectors has hampered the study of the fungi per se and of their involvement as vectors. None of them has been obtained in either pure or axenic culture; thus, evidence for their role as vectors has usually depended on indirect correlations between fungal infection and virus infection. Nevertheless, unifungal, or even better, single-sporangial isolates have been developed for several of these fungi. Such cultures are preferred for transmission studies. They provide a high degree of assurance as to the identity of the vector and its genetic homogeneity so that physiological specialization can be evaluated.

2.3 Relationships Between Virus and Vector

The two types of fungal vector—virus relationships were established from comparative studies of the transmission of tobacco necrosis virus (TNV) and the lettuce big-vein agent (BVA) by *Olpidium brassicae* (Campbell and Fry, 1966). Unfortunately, the failure to characterize the BVA, the lack of sap transmissibility of the BVA, and the lack of a local-lesion host have precluded several types of experiments. Subsequent studies with soil-borne wheat mosaic virus (SBWMV) and *Polymyxa graminis* have included such experiments and have verified the characteristics of this type of relationship (Rao and Brakke, 1969) (see Sect. 3.4.1).

The virus—vector relationships should be characterized first by the location of the virus relative to the fungal resting spore and second by the method of acquisition. The virus may be carried internally by the resting spores or be external to them. This can be determined quickly by treating resting spores with strong chemicals that can be shown to inactivate virus exposed to them. Thus, a virus carried internally will continue to be transmitted after treatment, whereas the external virus will not. It is presumed that the internally borne virus is also internally borne in the zoospores and in the vegetative thallus but there is no unequivocal test for this. Only one virus has been resolved within the vector (see Sect. 3.5). In addition, long-term survival of viral transmission in air-dry soil probably is due to virus within resting spores but such a test takes a long time and may not always be reliable (Smith et al., 1969). The method of virus acquisition by nonviruliferous isolates is correlated with the location of the virus. A virus carried internally is acquired by the fungus while it grows and reproduces in a virus-infected plant, i.e., in vivo acquisition. A virus that is external to the fungal resting spore is acquired by zoospores as they swim in a virus solution. Initially this was studied by mixing zoospores and virus in vitro. For convenience, we continue to refer to this as in vitro acquisition, but recognize that in nature it occurs in the soil water after virus and zoospores are released from host roots independently of each other (Smith et al., 1969). This acquisition probably is due to protein—protein interaction between the virus and protein components of the zoospore ectoplast. Protein conformations that permit this type of association are an additional requirement for a vector that acquires virus in vitro. A virus that is acquired in vitro and is external to the resting spores may well be taken into the zoospore protoplast during infection of the host.

There are some important implications of these relationships for the epidemiology of these viruses. The viruses carried internally, which constitute most of the known examples, are dependent on the vector both for their survival and their transmission. It is unlikely that the transmission of these viruses ever involves virus exposed outside the vector or host plant. Any virus that is liberated from living roots or from residue would be of no consequence except in experimental conditions where it could be detected by mechanical transmission to a host. On the other hand, the viruses that are external to the vector resting spores are unique among plant viruses in two respects. First, their survival in the absence of a crop or weed host results from their stability and ability to remain infective either in decomposing tissue or free in the soil. If these viruses are added to soil, they are rapidly inactivated by air-drying the soil (Smith et al., 1969). Thus, it is not surprising that these viruses are of importance in glasshouses or in agricultural soils, such as in northern Europe, where the soil is moist all year (van Slogteren and Visscher, 1967; Mowat, 1970). Second, these viruses are acquired by the vectors from the soil solution rather than by the vector feeding on or growing in an infected plant as occurs in all other virus—vector combinations.

3 Fungal Vectors and Viruses Transmitted by Them

3.1 Olpidium brassicae

This cosmopolitan fungus has been amenable to vector studies because it can be maintained in the vegetative or zoosporangial stage indefinitely, the vegetative stage is completed in about three days, and millions of zoospores can be produced on demand. With such characteristics, it is not surprising that it was the first fungal vector to be identified.

The ultrastructure of the fungus and host-parasite relationships have been studied in several laboratories (Temmink and Campbell, 1968, 1969a, b; Lesemann and Fuchs, 1970; Lange and Olson, 1976; Barr and Hartmann, 1977).

Single-sporangial isolates have been obtained and tested for physiological specialization (Sahtiyanci, 1962; Lin et al., 1970; Temmink et al., 1970). There are probably as many physiologically specialized isolates differing in infectivity to host species as one would care to test. Those tested, however, have not had the narrow host specificity of formae speciales or physiological races.

The effect of environmental factors on the fungus has been studied quantitatively. The most rapid maturation of zoosporangia is at 22°C, but the greatest numbers of zoospores are produced at 16°C (Fry and Campbell, 1966). Germination of resting spores responds similarly (Westerlund et al., 1978a). Soil moisture is more restrictive than temperature (Westerlund et al., 1978b). Resting spores will not germinate unless the matric potential of soil water is above − 60 millibars and zoospores are not released from sporangia in host roots until the matric potential is raised to 0 millibars (saturated soil).

3.1.1 Transmission of Lettuce Big-Vein Agent (BVA)

When *O. brassicae* was associated with lettuce big vein (Fry, 1958; Grogan et al., 1958), there was no evidence for a virus associated with the disease. Subsequently, the graft transmissibility of the disease was shown (Campbell et al., 1961; Tomlinson et al., 1962) and it was obvious that *O. brassicae* must be functioning as a vector. This was quickly confirmed by tests correlating *O. brassicae* infection and BVA transmission (Campbell and Grogan, 1963; Tomlinson and Garrett, 1964) and later single-sporangial isolates were used to prove that *O. brassicae* was the vector (Lin et al., 1970). The BVA was shown to be protected within resting spores (Campbell, 1962; Cambell and Fry, 1966). In vivo acquisition by nonviruliferous cultures was demonstrated (Tomlinson and Garrett, 1962) and it was shown to occur during one generation (Campbell and Grogan, 1964). Transmission of BVA to the host plant also seems to occur, at least at a low rate, during the first generation that *O. brassicae* is in the root system (Westerlund et al., 1978a). The BVA is not uniformly distributed among the vector thalli, and presumably not among the epidermal cells of the host root either, because only about half of the single-sporangial isolates from a bulk culture transmitted BVA (Lin et al., 1970). Finally, the BVA does not seem to multiply in the vector because viruliferous cultures can be freed of BVA by culture on sugar beets for 14 weeks or longer (Campbell, 1962) or on *Plantago major* for between 6 and 52 weeks (Tomlinson and Garrett, 1964).

The nature of the BVA has not been determined. For a time it seemed likely that a virus could be demonstrated and it was called the big-vein virus. Attempts to characterize the presumed virus have not succeeded and it now seems best to call it the BVA. Big-vein agent is not mechanically transmitted, no viruslike or mycoplasmalike or rickettsialike structures have been seen in *O. brassicae* or the host, and no viroidlike RNA has been detected (Campbell and Grogan, 1963, 1964; Tomlinson and Garrett, 1964; Lin et al., 1970; Westerlund et al., 1978b). There is a brief report of mycoplasmalike, rickettsialike, and viruslike particles in plants with big vein (Ragozzino and Furia, 1972) but this has not been confirmed. A 244-nm rod-shaped particle has been associated with the roots of big-vein plants (Chod et al., 1976). We have attempted to verify this report and found the same type of rods in plants with BVA-transmitting *O. brassicae* but also in the controls that had BVA-free *O. brassicae* (Lecoq and Campbell, unpublished). The BVA may be a virus that has escaped detection or a unique agent but, until a satisfactory method of assaying for it in extracts has been developed, it is likely to remain uncharacterized.

3.1.2 Transmission of Tobacco Necrosis Virus (TNV)

TNV is a well-characterized virus that causes local lesions on many hosts but seldom infects systemically. It has polyhedral particles about 26 nm in diameter and there are many strains that are serologically related.

Some correlations between fungal and virus infection were established (Teakle, 1962), but some experiments were inconclusive. Later, additional correlations were established (Fry and Campbell, 1966) and, finally, single-sporangial cultures were shown to transmit TNV (Temmink et al., 1970).

The relationship between zoospores and TNV was studied using indirect tests or materials that were not adequate to discriminate between the hypotheses being tested. Although in vitro acquisition was used for most trials, the virus was claimed to be within zoospores (Teakle and Gold, 1963) or loosely bound to zoospores (Kassanis and MacFarlane, 1964). Later, other indirect tests supported the hypothesis of a tight bind-

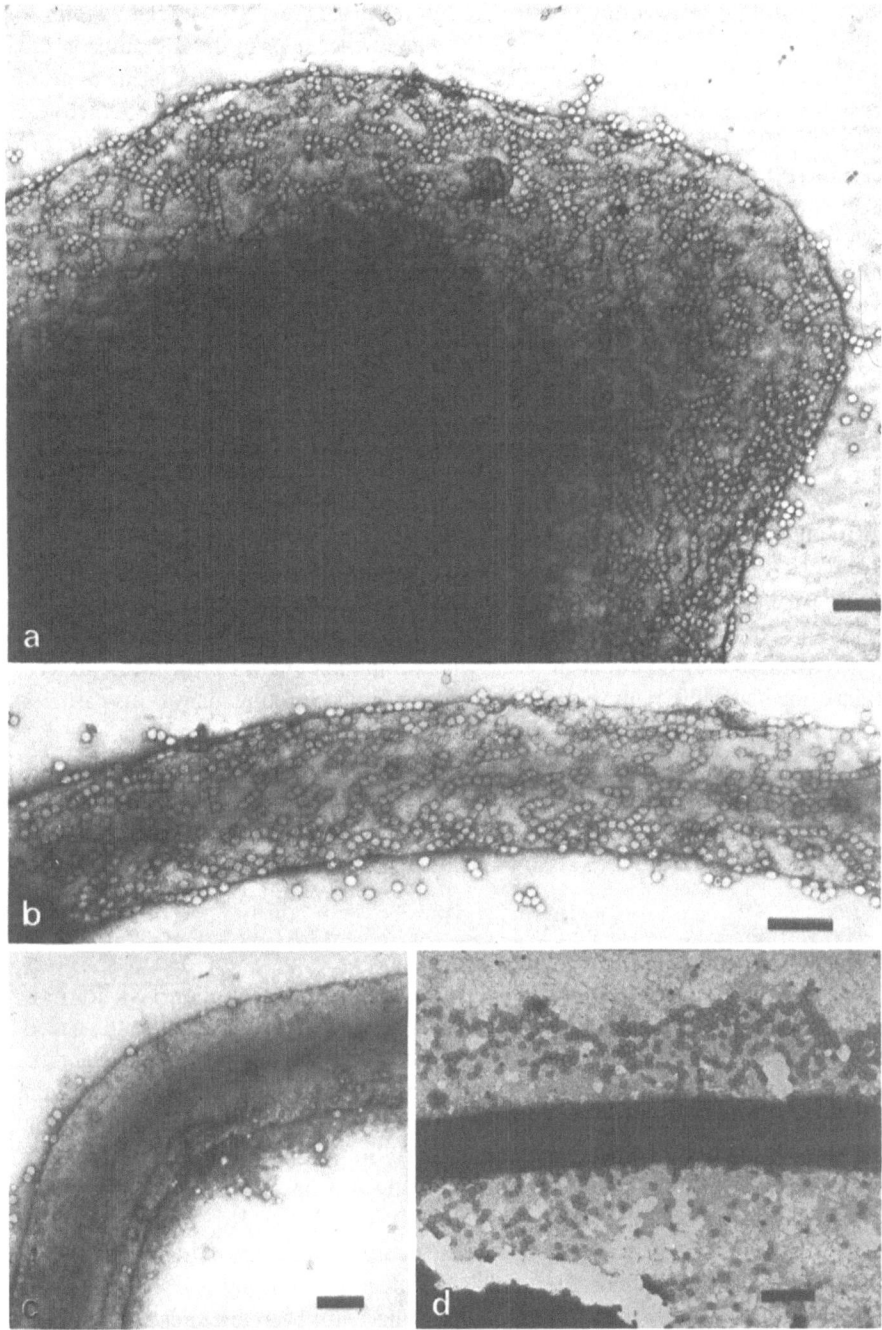

Fig. 1a–d. Zoospores of *Olpidium* spp. with virus acquired in vitro. In each photograph zoospores and virus were mixed and the excess or nonacquired virus removed by washing the zoospores before fixation for electron microscopy. **a, b** *Olpidium brassicae* with tobacco necrosis virus (TNV) on the ectoplast of the "head" and flagellum. **c** *O. brassicae* flagellum with both TNV (26 nm particles) and satellite virus (17 nm particles). **d** *O. radicale* flagellum with cucumber necrosis virus. Note that more virus had dissociated from the flagellum during fixation than occurred with TNV in **b** *Bar* = 200 nm

ing of virus to the zoospore ectoplast (Campbell and Fry, 1966). This has been confirmed by electron microscopy of intact zoospores that have TNV particles adhering to them (Fig. 1a, b) (Temmink et al., 1970). The further steps in the transmission process doubtless involve movement of virus into the zoospore protoplast during or after encystment and its release after the zoospore protoplast has infected the host cell (Temmink, 1971). The relative fragility of the zoospores can confound the results of experiments to determine the virus—zoospore relationship as shown above. For this reason, the virus-resting spore relationship is easier to determine and is preferred. TNV is external to the resting spores (Campbell and Fry, 1966) as well as to thalli developing in virus-infected cells (Temmink, 1971). Necrosis of host cells from TNV infection can have an adverse effect on *O. brassicae* multiplication because the necrosis causes death of the zoosporangia before they mature (Fry and Campbell, 1966). On the other hand, TNV transmission does not require that *O. brassicae* do more than infect the host cells (Kassanis and MacFarlane, 1965). This situation apparently occurs in nature in tulips which may develop Augusta disease from systemic infection by TNV but which never allow development of the thalli (Mowat, 1970; van Slogteren and Visscher, 1967). Other viruses not transmitted by *O. brassicae* were not acquired by zoospores in in vitro mixtures (Temmink et al., 1970).

There is a wide variation in the vector efficiency of different isolates of *O. brassicae*. Isolates of *O. brassicae* from crucifers have not transmitted any of several strains of TNV (Teakle and Hiruki, 1964; Kassanis and MacFarlane, 1965; Mowat, 1968) apparently because they do not acquire the virus (Temmink et al., 1970). At the other extreme, the highly efficient vectors used in most studies acquire many virus particles and transmit most, if not all, strains of TNV. In between are a group of isolates (e.g., the oat isolate of Temmink et al., 1970) that transmit some strains of TNV to some hosts, but not all strains to all hosts. In these cases the specificity may be due to inefficient acquisition, to failure of later stages of transmission after fungal penetration, or to responses of the host cells (Kassanis and MacFarlane, 1965; Temmink et al., 1970).

3.1.3 Transmission of Satellite Virus (SV)

SV is a small (17 nm diameter) polyhedral virus that is dependent upon TNV for its multiplication and occurs as several serologically related strains. Because of the dependence upon TNV one would suspect SV would be transmitted in the same manner as TNV. Kassanis and MacFarlane (1968) mixed zoospores of *O. brassicae* with TNV and SV in vitro and demonstrated that both viruses were transmitted to and multiplied in host plants. Isolates of *O. brassicae* varied in ability to transmit different isolates of SV. Electron microscopic examination of zoospores showed that SV and TNV were both acquired and tightly bound to the zoospore plasmalemma (Fig. 1c) (Temmink et al., 1970). This means that on the ectoplast of the zoospore there are binding sites for at least two distinct viral proteins. Presumably, SV does not multiply in *O. brassicae* and is external to the resting spores, but these points have not been tested experimentally.

3.1.4 Transmission of Tobacco Stunt Agent (TSA)

Tobacco stunt disease has been much studied but unfortunately much of the evidence has only been presented as abstracts without details of methodology and results.

The TSA is reported to be mechanically transmissible but it has not yet been satis-factorily characterized. Hidaka (1956) extracted a 25 nm spherical particle from dis-eased tobacco. Later, 18 nm particles were found in ultrathin sections of tobacco and *O. brassicae* (Fukishima and Hidaka, 1969). Another report from the same laboratory avoided the use of "virus," and concluded that the exact nature of the TSA was un-known (Hidaka et al., 1975). Nevertheless, their descriptions of such characteristics as a dilution-end-point between 10^{-2} and 10^{-3} and a longevity in vitro at $4°C$ between 6 and 16 h suggest a labile virus might be involved.

The only other work on tobacco stunt disease was done by Hiruki who refers to the causal agent as the tobacco stunt virus (Hiruki, 1975). A 40–60 nm particle was extracted from diseased tobacco tissue (Hiruki et al., 1975), but it has not been proved that this is the causal agent. This virus has a longevity in vitro of 60–84 h at $20°C$, and a dilution-end-point between 10^{-2} to 10^{-3} if 1-phenyl-thiosemicarbazide is added to the extraction buffer (Hiruki, 1975). This virus has a wide host range by mechanical transmission and the host range by *O. brassicae* transmission is not the same (Hiruki, 1967, 1975). Apparently no test has been made to confirm that this virus is transmitted by *O. brassicae* after mechanical transmission to and recovery from hosts outside the Solanaceae.

A number of correlations have been established to support the hypothesis that *O. brassicae* is the vector (Hidaka and Tagawa, 1962, 1965; Hiruki, 1965), but no tests have been made with single-sporangial isolates. The TSA seems to be internally borne because transmission was reported to occur after treatment of resting spores (Hiruki, 1972). In vivo acquisition from mechanically inoculated tobacco plants (Hiruki, 1965, 1972; Hidaka et al., 1975) or from graft-inoculated tobacco plants (Hidaka and Taga-wa, 1965) has been reported. The purported occurrence of crystals of virus in zoo-spores would suggest that TSA multiplies in the vector (Fukishima and Hidaka, 1969) but this evidence has not been published. The putative zoospore with a virus crystal shown in an electron micrograph (Z. Hidaka, personal communication) has little re-semblance to zoospores of *O. brassicae* in published papers (Temmink and Campbell, 1969a; Lange and Olson, 1976; Barr and Hartmann, 1977). Alternatively, support for the hypothesis of TSA multiplication in the fungus could be obtained if the fungus culture maintained TSA while cultured on nonhosts of TSA. The experimental evi-dence from such trials is conflicting and inconclusive. Hidaka et al. (1975) reported that *O. brassicae* lost the TSA when kept for five transfers at 20-day intervals on lettuce, but Hiruki (1967) reported lettuce was a host for TSA transmitted by *O. brassicae*. On the other hand, Hiruki (1965) reported his isolate lost TSA during an unspecified time on cowpea although the leaves of this host were susceptible by mechanical transmis-sion (Hiruki, 1967, 1975).

3.2 Olpidium radicale Transmission of Cucumber Necrosis Virus (CNV)

Olpidium radicale is similar to *O. brassicae* but has a larger zoospore and a smooth-walled resting spore; *O. cucurbitacearum* has been reduced to synonymy with *O. radi-cale* (Lange and Insunza, 1977). CNV is a polyhedral virus about 28 nm in diameter but serologically distinct from strains of TNV.

Dias (1970b) provided several correlations to show that *O. radicale* was the vector of CNV. He also showed that CNV is external to the resting spore and that zoospores of *O. radicale* acquire CNV, but not TNV, in vitro (Dias, 1970a). The in vitro acquisition of CNV, but not of TNV, was confirmed by electron microscopy of *O. radicale* zoospores (Temmink et al., 1970). These authors observed that CNV dissociated more easily from zoospores of the vector (Fig. 1d) than did TNV from its vector.

3.3 Synchytrium endobioticum and Transmission of Potato Virus-X (PV-X)

Synchytrium endobioticum causes potato wart disease and is subject to quarantine in many countries. PV-X is a flexuous rod-shaped particle about 515 nm long and is easily transmitted mechanically. There is no necessity for a vector when the virus survives in a perennial host (potato) and can be readily spread by contact between plants in the field. Nevertheless, it seems likely that a vector must have been of importance for the dissemination of the virus prior to the advent of agriculture or in annual hosts. Because agriculture is a recent development on the time scale of evolution, it is likely that a vector might still be active in some places. The report that *S. endobioticum* acquired PV-X in vivo, but not in vitro, offers evidence of fungal transmission (Nienhaus and Stille, 1965). Presumably the virus is carried internally in the resting spores, but this has not been tested. Unfortunately, there has been no further research to confirm or extend these initial observations.

3.4 Polymyxa graminis

This fungus is distributed worldwide in the roots of cereals. It has not been extensively studied because (1) it is a parasite, not a pathogen, and (2) it is a difficult fungus to manage. Detailed, light microscopic observations have been made by Ledingham (1939) and Rao (1968) but no ultrastructural studies have been reported. The fungus completes a vegetative life cycle in 8—9 days and releases zoospores; few zoospores are seen at any time (Ledingham, 1939; Rao, 1968; Slykhuis and Barr, 1978). Only recently have unifungal cultures been developed (Slykhuis and Barr, 1978).

3.4.1 Transmission of Soil-Borne Wheat Mosaic Virus (SBWMV)

SBWMV has stiff rods about 20 nm in diameter and of two modal lengths (110—160 and about 300 nm). The length of the shorter rods differs with the isolate. The virus is known from the USA, Europe and Japan. Serologically, it is distantly related to tobacco mosaic virus (Powell, 1976; Randles et al., 1976) and more closely related to potato mop-top virus (Randles et al., 1976).

SBWMV was clearly established as a soil-transmitted virus by McKinney and coworkers (McKinney et al., 1957). They observed *P. graminis* in infected plants but the correlation was imperfect because no fungus was found in some of the mosaic-infected plants (Linford and McKinney, 1954). They clearly suspected *P. graminis* and continued experimentation, but were unable to obtain conclusive results (unpublished notes cited in Brakke et al.,1965). The development of correlations between *P. graminis* and virus transmission (Estes and Brakke, 1966) did not occur until better methods of inoculation and assay were developed (Brakke et al., 1965; Brakke and Estes, 1967). Canova (1966) developed similar evidence in Italy at about the same time.

Many attributes of the virus—vector relationship have been established: (1) the virus is internally borne in resting spores, (2) virus is acquired in vivo but not in vitro, (3) apparently the virus is within zoospores because SBWMV-antiserum mixed with zoospores did not prevent transmission, (4) the virus is released into healthy hosts between 6 and 22 h after inoculation with zoospores (Rao and Brakke, 1969). The virus apparently does not multiply in the fungus because nonviruliferous cultures were obtained after three passages of unspecified time periods in clover (Canova, 1966). Virus particles have not been resolved within the fungus (W.G. Langenberg, personal communication).

3.4.2 Transmission of Flexuous Rod-Shaped Viruses of Cereals

Five viruses with similar characteristics have been reported from cereals and will be considered as a group. They all infect cereals but differ in host range and reaction; have similar size particles, 13 nm in diameter, but of different lengths; cause formation of pinwheel type inclusions in their hosts; and are soil-transmitted (Slykhuis, 1976; Inouye and Fujii, 1977). Barely yellow mosaic virus (BYMV), wheat yellow mosaic virus (WYMV), and rice necrosis virus (RNV) occur in Japan, are serologically related and have two modal lengths of particles, 275 and 550 nm (Inouye and Saito, 1975; Inouye and Fujii, 1977). Oat mosaic virus (OMV) that occurs in the USA and United Kingdom has particles 600—750 nm long (Hebert and Panizo, 1975). Wheat spindle streak virus (WSSV) that occurs in the USA and France has particles from 190—1975 nm associated with it (Slykhuis, 1976). The viruses are difficult to purify; aggregation or fragmentation of particles may account for the wide range of particle lengths.

Correlations between *P. graminis* infections and virus transmission have been reported in each of the references cited above. In vivo acquisition has been demonstrated for BYMV (Toyama and Kusaba, 1970) and WSSV (Slykhuis and Barr, 1978). In the latter work the authors used unifungal cultures to prove *P. graminis* was a vector, but four other lower fungi were not. Presumably these viruses are all internally borne in the resting spores of the fungal vector as has been reported for BYMV by Kusaba et al. (1971) as cited by Inouye and Saito (1975).

3.5 Polymyxa betae and Transmission of Beet Necrotic Yellow Vein Virus (BNYVV)

The second species of *Polymyxa, P. betae*, was described in 1964; its host range is limited to Chenopodiaceae and it causes stunting of sugar beet plants (Keskin, 1964). BNYVV, first named in 1973, has particles that are 20 nm in diameter and fall into two or three length groups; these depend on the isolate and are 60—105 nm, 270 nm and 390 nm (Tamada, 1975). No serological relationships have been shown between BNYVV and TMV, SBWMV, potato mop-top virus, peanut clump virus, or *Nicotiana velutina* mosaic virus (Tamada, 1975; Randles et al., 1976; Thouvenel et al., 1976; Kuszala and Putz, 1977). BNYVV by itself seems to be the primary causal agent of the rhizomania disease of sugar beets (Tamada, 1975) in Japan and Europe.

Polymyxa betae infection and transmission of BNYVV were correlated (Tamada et al., 1975; Vuittenez et al., 1977) and virus-free isolates acquired BNYVV in vivo (Tamada et al., 1975). No single-sporangial or unifungal cultures have been tested and treatments of resting spores have not been done; presumably the virus is internally borne in resting spores. This is strongly indicated by the report that viruslike rods have been observed in the fungal zoospore (Tamada, 1975; Stocky et al., 1977) and in the plasmodial thallus (Stocky et al., 1977). To date only one electron micrograph has been published (Tamada, 1975) representing the first visual detection of virus particles within a fungal vector. Of interest, but not known, is whether these particles represent virus multiplication in the vector or acquisition after synthesis in the host.

3.6 Spongospora subterranea and Transmission of Potato Mop-Top Virus (PMTV)

This fungus is the cause of powdery scab of potatoes. PMTV occurs in Europe and in South America, has particles that are 20 nm × 100–150 and 250–300 nm (Kassanis et al., 1972), and is serologically related to SBWMV (Randles et al., 1976) and to TMV (Kassanis et al., 1972).

Correlations between fungal and virus infections have been established (Jones and Harrison, 1969) but no other information on fungus–virus relationships has been published. The survival of virus transmission when fungal resting spores were dried for one year before testing suggests, but does not prove, that PMTV is internally borne in the resting spores.

4   Soil-Borne Viruses That May Have Fungal Vectors

4.1 Tobacco Mosaic Virus (TMV)

Several viruses have been listed as tentative members of the tobacco mosaic virus group (Gibbs, 1977), these include SBWMV, BNYVV, and PMTV whose serological relationships to TMV and whose fungal transmission has been discussed. There are other soil-borne viruses in this tentative group: (1) peanut clump virus (Thouvenel et al., 1976) that has rod-shaped particles 190 and 245 nm long but is not serologically related to tobacco rattle virus, pea early browning virus, SBWMV, or BNYVV; (2) broadbean necrosis virus that has rod-shaped particles 150 and 250 nm long (Inouye and Asatani, 1968); (3) oat tubular virus that has particles 150 and 305–330 nm long (Plumb et al., 1977). It seems quite likely that a fungal vector exists for these viruses; a plasmodiophoromycete vector seems the most likely.

In contrast, there are a number of definitive members of the tobacco mosaic virus group (Gibbs, 1977). These have no recognized vector and little need for a vector in the conditions of agriculture. Their stability and occasional seed (embryonic infection or seed-coat infestation) transmission permit survival and primary infection. Epidemics develop as they spread in their respective crop plants by methods that are basically sap transmission. Nevertheless, as argued for PV-X above, it seems quite possible that these viruses had a vector(s) until recent time on an evolutionary scale, that the vector could have been a fungus, and that this vector still functions somewhere today but without impact on cultivated plants. For that matter, the presumed vector could be active in transmission of TMV to plants other than tobacco or tomato. While the virus could have evolved to a condition in which it is not acquired or transmitted by its vector during the short interval from the beginning of agriculture, the primitive vector—virus relationship should still exist and would constitute an interesting system to study.

Some attempts to demonstrate fungal transmission of TMV have been reported. A powdery mildew was claimed to transmit a TMV strain from oak trees to *Chenopodium* sp. (Yarwood, 1971), but the evidence was inconclusive (Yarwood and Hecht-Poinar, 1973) and could have been explained by other mechanisms. There is good evidence that TMV can be inoculated into and multiply in some *Pythium* spp. (Brandts, 1969, 1971; Nienhaus and Mack, 1974; Pahlow, 1976). This extends the host range of TMV, and makes it a subject for other authors in this symposium, but does not prove or disprove the transmission of TMV by *Pythium* spp. to plant hosts.

4.2 Tomato Bushy Stunt Virus (TBSV) Group

This large group of polyhedral viruses has no known vector. One member of the group, *Petunia* asteroid mosaic virus (PAMV), however, has been reported to be soil-transmitted (Lovisolo et al., 1965). Campbell et al. (1975) established some characteristics of the presumed vector that seemed to be a zoospore-producing fungus, but were unable to isolate or identify it. Neither *O. brassicae* nor *Lagena radicicola* Vant & Led. were the vectors. [Barr has indicated that the binomial *L. radicicola* should be continued in preference to *Lagenocystis radicicola* (Vant. & Led.) Copeland that was used in Campbell et al. (1975) (D.J.S. Barr, personal communication).]

The epidemiology of TBSV is not clear. Some isolates in perennial hosts (*Pelargonium*, artichoke, carnation) are occasionally mechanically transmitted or else are propagated when infected hosts are vegetatively propagated. Other isolates that occur sporadically in annual crops may also result from mechanical sap transmission from an unidentified reservoir or may result from an erratic distribution and activity on the part of a fungal vector. One would expect sooner or later to see an epidemic in a field if soil transmission was active or efficient.

Cymbidium ringspot virus (Cy RSV) has been provisionally assigned to the TBSV group although it has no serological affinities with members of the group (Hollings et al., 1977). Cy RSV was said to be soil-transmitted but there was no evidence of vector activity. In the terminology of the present paper it would be classified as "soil-borne under experimental conditions."

4.3 Miscellaneous

A number of other viruses have been reported to be soil-borne or soil-transmitted. Those suspected of having a fungal vector will be reviewed.

Freesia leaf necrosis was shown to be a soil-borne agent and a fungal vector was suggested (van Dorst, 1975). These preliminary experiments need to be continued to determine both the nature of the causal agent and its soil transmissibility.

Red clover necrotic mosaic virus (RCNMV) has been described as a virus that can be soil-borne (Hollings and Stone, 1977). Lange and Insunza (1977) suggest that *O. radicale* may be the vector. Both the virus and *O. radicale* are easily managed, so further results should be forthcoming. If *O. radicale* is the vector, comparisons between RC-NMV and CNV would be fruitful as they have similar characteristics and the same vector.

Melon necrotic spot virus was suspected to be soil-borne and to be transmitted by *Olpidium* sp. (Komuro, 1972). Its relationship to RCNMV or to CNV needs to be determined as well as the specific identity of the vector and the virus—vector relationship.

Sugar cane mosaic virus, that is readily transmitted by aphids, apparently can be soil-borne under experimental conditions (Bond and Pirone, 1970). There was no evidence that a vector was present in the soil or that there was root contact. This may be a case where virus released into the soil from an infected plant is later mechanically inoculated to the roots of a healthy plant by abrasion. If so, it would indicate that the virus can persist in sterilized soil longer than expected. These observations should be tested experimentally.

Sugar cane chlorotic streak remains a perplexing disease. Since the review in 1966 (Grogan and Campbell, 1966), a few facts have been added. The chlorotic streak agent (CSA) is inactivated in vivo in 1/2 h at 50°C compared to the ratoon stunting agent that requires 3 h at 50°C, and it is not affected by four tetracycline antibiotics (Anonymous, 1972). The rapid heat inactivation of the agent is more characteristic of mycoplasmalike agents than of viruses, but insensitivity to antibiotics is not. As discussed previously (Grogan and Campbell, 1966), it is difficult to imagine that a soil-infesting nematode or fungal vector is involved when stem cuttings can be rooted and serve as sources of soil-borne inoculum.

Pea false leaf roll virus is of interest because it has been described as being seed-, aphid-, and *Pythium ultimum*-transmitted (Thottappilly and Schmutterer, 1968; Thottappilly, 1972). Transmission by both an aphid and a fungus would be unique as would transmission by *P. ultimum*. While the experiments reported by these authors support the hypothesis that PFLRV is transmitted by *P. ultimum*, we lack some critical tests of this hypothesis. First, the morphology of the PFLRV needs to be established. It is mechanically transmissible and this should not be difficult to do. Second, PFLRV disease must be shown to be caused by a single virus and not a complex, for example, with pea seed-borne mosaic virus. The disease symptoms should be reproducible by a pure culture of PFLRV and should be clearly distinguished from symptoms of *Pythium* root rot or pea seed-borne mosaic virus. Third, the pure culture of PFLRV should be shown to be aphid- and fungus-transmitted in successive trials. So far, no tests of in vivo or in vitro acquisition with *P. ultimum* have been reported. Finally, loss of PFLRV from the fungal vector should be demonstrable either in pure culture or on nonhosts of the virus.

Summary

Soil-borne virus diseases have been known since 1919 but only in the last 20 years have the existence and identity of the vectors been demonstrated. These vectors may be nematodes or fungi but are specific for each known virus. This paper will review the nature of virus—fungal vector relationship and summarize the viruses known to have fungal vectors.

Two basic types of fungus—virus relationship have been established. In the first the virus and fungal zoospores are released independently from the infected host roots. The fungus acquires the virus in vitro or in soil water by adsorption to the plasma membrane of the zoospore. The precise method by which adsorbed virus is introduced into the host is uncertain but presumably the virus is taken into the zoospore protoplasm during encystment or infection. The virus does not occur in the fungus thallus as it grows and forms zoospores or resting spores. Thus, the virus must be capable of survival from season to season in infected roots or root residues or free in the soil. Tobacco necrosis virus and *Olpidium brassicae* illustrate this relationship.

In the second type of relationship acquisition of virus by nonviruliferous vectors occurs only in vivo in the roots of an infected host. The virus is apparently internal in the zoospores as they are released from the infected host. The virus presumably is carried in the zoospore protoplast as it infects the host cell and somehow is released into the host cell where it multiplies. The virus occurs within the thallus as zoospores or resting spores are formed. Thus, the virus is protected within the fungus resting spores for survival from season to season or crop to crop. An example is soil-borne wheat mosaic virus and *Polymyxa graminis*.

The known vectors are in the Plasmodiophoromycetes (*Polymyxa* spp. and *Spongospora* sp.) and the Chytridiales (*Olpidium* spp. and *Synchytrium* sp.). All produce thick-walled, soil-borne resting spores, as well as zoospores, and are obligate parasites. The viruses range from flexuous rods to stiff rods to isometric in shape. The known virus–vector combinations are tabulated according to the type of relationship.

References

Anonymous Annu Report (1972) Mauritius sugar industry research institute, Port Louis, Mauritius, (RPP) 54:981b

Barr DJS, Hartmann VE (1977) Zoospore ultrastructure of *Olpidium brassicae* and *Rhizophlyctis rosea*. Can J Bot 55:1221–1235

Bond WP, Pirone TP (1970) Evidence for soil transmission of sugarcane mosaic virus. Phytopathology 60:437–440

Brakke MK, Estes AP (1967) Some factors affecting vector transmission of soil-borne wheat mosaic virus from root washings and soil debris. Phytopathology 57:905–910

Brakke MK, Estes AP, Schuster ML (1965) Transmission of soil-borne wheat mosaic virus. Phytopathology 55:79–86

Brandts DH (1969) Tobacco mosaic virus in *Pythium* spec. Neth J Plant Pathol 75:296–299

Brandts DH (1971) Infection of *Pythium sylvaticum* in vitro with tobacco mosaic virus. Neth J Plant Pathol 77:175–177

Campbell RN (1962) Relationship between the lettuce big-vein virus and its vector, *Olpidium brassicae*. Nature (London) 195:675–677

Campbell RN, Fry PR (1966) The nature of the associations between *Olpidum brassicae* and lettuce big-vein and tobacco necrosis viruses. Virology 29:222–233

Campbell RN, Grogan RG (1963) Big-vein virus of lettuce and its transmission by *Olpidum brassicae*. Phytopathology 53:252–259

Campbell RN, Grogan RG (1964) Acquisition and transmission of lettuce big-vein virus by *Olpidium brassicae*. Phytopathology 54:681–690

Campbell RN, Grogan RG, Purcifull DE (1961) Graft transmission of big vein of lettuce. Virology 15:82–85

Campbell RN, Lovisolo O, Lisa V (1975) Soil transmission of *Petunia* asteroid mosaic strain of tomato bushy stunt virus. Phytopathol. Mediterr 14:82–86

Canova A (1966) Ricerche sulle malattie da virus delle Graminacee III. *Polymyxa graminis* Led. vettore del virus del mosaico del Frumento. Phytopathol Mediterr 5:53–58

Chod J, Polák J, Kůdela V, Jokeš M (1976) Finding of lettuce big vein virus in Czechoslovakia. Biol Plant 18:63–66

Dias HF (1970a) The relationship between cucumber necrosis virus and its vector, *Olpidium cucurbitacearum*. Virology 42:204–211

Dias HF (1970b) Transmission of cucumber necrosis virus by *Olpidium cucurbitacearum* Barr & Dias. Virology 40:828–839

Dorst HJM van (1975) Evidence for a soilborne nature of *Freesia* leaf necrosis. Neth J Plant Pathol 81:45–48

Estes AP, Brakke MK (1966) Correlation of *Polymyxa graminis* with transmission of soil-borne wheat mosaic virus. Virology 28:772–774

Fry PR (1958) The relationship of *Olpidium brassicae* (Wor.) Dang. to the big-vein disease of lettuce. N Z J Agric Res 1:301–304

Fry PR, Campbell RN (1966) Transmission of a tobacco necrosis virus by *Olpidium brassicae.* Virology 30:517–527

Fukishima Y., Hidaka Z (1969) Observations on the virus particles in the TSV-infected tissues and in its vector *Olpidium brassicae.* Ann Phytopathol Soc Japan (Abstr) 35:124

Gibbs AJ (1977) Tobamovirus group. CMI/AAB descriptions of plant viruses no 184, pp 6

Grogan RG, Campbell RN (1966) Fungi as vectors and hosts of viruses. Annu Rev Phytopathol 4: 29–52

Grogan RG, Zink FW, Hewitt WB, Kimble KA (1958) The association of *Olpidium* with the big-vein disease of lettuce. Phytopathology 48:292–296

Hebert TT, Panizo CH (1975) Oat mosaic virus. CMI/AAB descriptions of plant viruses no 145, pp 4

Hewitt WB, Raski DJ, Goheen AC (1958) Nematode vector of soil-borne fanleaf virus of grapevines. Phytopathology 48:586–595

Hidaka S, Tagawa H, Hidaka Z (1975) Acquisition and loss of the tobacco stunt pathogen by *Olpidium brassicae,* and transmission by sap inoculation. Proc 1st Intersect Congr Int Assoc Microbiol Soc 3:287–296

Hidaka Z (1956) Morphology of the tobacco stunt virus. Proc 3rd Int Conf Electron Microscopy London 1954. R Microsc Soc London, pp 256

Hidaka Z, Tagawa H (1962) The relationship between the occurrence of tobacco stunt disease and *Olpidium brassicae.* Ann Phytopathol Soc Jpn 27:77–78 (abstr)

Hidaka Z, Tagawa H (1965) Transmission of tobacco stunt virus by *Olpidium brassicae* (Wor.) Dang. Ann Phypathol Soc Jpn 31:369–372

Hiruki C (1965) Transmission of tobacco stunt virus by *Olpidium brassicae.* Virology 25:541–549

Hiruki C (1967) Host specificity in transmission of tobacco stunt virus by *Olpidium brassicae.* Virology 33:131–136

Hiruki C (1972) Persistence of tobacco stunt virus in resting sporangia of *Olpidium brassicae.* Int Virol 2:249 (abstr)

Hiruki C (1975) Host range and properties of tobacco stunt virus. Can J Bot 53:2425–2434

Hiruki C, Alderson PG, Kobayashi N, Furusawa I (1975) The nature of the infectious agent of tobacco stunt in relation to its vector, *Olpidium brassicae.* Proc 1st Intersect Congr Int Assoc Microbiol Soc 3:297–302

Hollings M, Stone OM (1977) Red clover necrotic mosaic virus CMI/AAB descriptions of plant viruses no 181, p 4

Hollings M, Stone OM, Barton RJ (1977) Pathology, soil transmission and characterization of *Cymbidium* ringspot, a virus from *Cymbidium* orchids and white clover *(Trifolium repens).* Ann Appl Biol 85:233–248

Inouye T, Asatani M (1968) Broadbean necrosis virus. Ann Phytopathol Soc Jpn 34:317–322

Inouye T, Fujii S (1977) Rice necrosis mosaic virus. CMI/AAB descriptions of plant viruses no 172, pp 4

Inouye T, Saito Y (1975) Barley yellow mosaic virus. CMI/AAB descriptions of plant viruses no 143, pp 3

Jones RAC, Harrison BD (1969) The behaviour of potato mop-top virus in soil and evidence for its transmission by *Spongospora subterranea* (Wallr.) Lagerh. Ann Appl Biol 63:1–17

Kassanis B, MacFarlane I (1964) Transmission of tobacco necrosis by zoospores of *Olpidium brassicae.* J Gen Microbiol 36:79–93

Kassanis B, MacFarlane I (1965) Interaction of virus strain, fungus isolate, and host species in the transmission of tobacco necrosis virus. Virology 26:603–612

Kassanis B, MacFarlane I (1968) The transmission of satellite viruses of tobacco necrosis by *Olpidium brassicae*. J Gen Virol 3:227–232

Kassanis B, Woods RD, White RF (1972) Some properties of potato mop-top virus and its serological relationship to tobacco mosaic virus. J Gen Virol 14:123–132

Keskin B (1964) *Polymyxa betae* n sp, ein Parasit in den Wurzeln von *Beta vulgaris* Tournefort, besonders während der Jugendentwicklung der Zuckerrübe. Arch Mikrobiol 49:348–374

Komuro Y (1971) Cucumber green mottle mosaic virus on cucumber and watermelon and melon necrotic spot virus on muskmelon. Jpn Agric Res Q 6:41–45 (RPP 51:3007, 1972)

Kuszala M, Putz C (1977) La Rhizomanie de la Betterave sucrière en Alsace. Gamme d'hôtes et propriétés biologiques du "Beet necrotic yellow vein virus." Ann Phytopathol 9:435–446

Lamberti F, Taylor CE, Seinhorst JW (eds) (1975) Nematode vectors of plant viruses, Vol 1. Plenum Press, New York, pp 460

Lange L, Insunza V (1977) Root-inhabiting *Olpidium* species: The *O. radicale* complex. Trans Br Mycol Soc 69:377–384

Lange L, Olson LW (1976) The zoospore of *Olpidium brassicae*. Protoplasma 90:33–45

Ledingham GA (1939) Studies on *Polymyxa graminis*, N Gen N Sp, a plasmodiophoraceous root parasite of wheat. Can J Res 17:38–51

Lesemann DE, Fuchs WH (1970) Die Ultrastruktur des Penetrationsvorganges von *Olpidium brassicae* an Kohlrabi-Wurzeln. Arch Mikrobiol 71:20–30

Lin MT, Campbell RN, Smith PR, Temmink JHM (1970) Lettuce big-vein virus transmission by single-sporangium isolates of *Olpidium brassicae*. Phytopathology 60:1630–1634

Linford MB, McKinney HH (1954) Occurrence of *Polymyxa graminis* in roots of small grains in the United States. Plant Dis Rep 38:711–713

Lovisolo O, Bode O, Völk J (1965) Preliminary studies on the soil transmission of *Petunia* asteroid mosaic virus (= "Petunia" strain of tomato bushy stunt virus). Phytopathol Z 53:323–342

McKinney HH, Paden WR, Koehler B (1957) Studies on chemical control and overseasoning of, and natural inoculation with, the soil-borne viruses of wheat and oats. Plant Dis Rep 41:256–266

Mowat WP (1968) *Olpidium brassicae:* Electrophoretic mobility of zoospores associated with their ability to transmit tobacco necrosis virus. Virology 34:565–568

Mowat WP (1970) Augusta disease in tulip – a reassessment. Ann Appl Biol 66:17–28

Nienhaus F, Mack C (1974) Infection of *Pythium arrhenomanes* in vitro with tobacco mosaic virus and tobacco necrosis virus. Z Pflanzenkr Pflanzenschutz 81:728–731

Nienhaus F, Stille B (1965) Übertragung des Kartoffel-X-Virus durch Zoosporen von *Synchytrium endobioticum*. Phytopathol Z 54:335–337

Pahlow G (1976) Translokation von Tabakmosaikvirus durch das Myzel kunstlich infizierter *Pythium*-Arten. Phytopathol Z 86:361–364

Plumb RT, Catherall PL, Chamberlain JA, Macfarlane I (1977) A new virus of oats in England and Wales. Ann Phytopathol 9:365–370

Powell CA (1976) The relationship between soil-borne wheat mosaic virus and tobacco mosaic virus. Virology 71:453–462

Ragozzino A, Furia A (1972) Osservazioni preliminari al microscopio elettronico di tessuti di lattuga affetta da ingrossamento nervale (big vein). Riv Patolog Veg 8:321–322

Randles JW, Harrison BD, Roberts IM (1976) *Nicotiana velutina* mosaic virus: purification, properties and affinities with other rod-shaped viruses. Ann Appl Biol 84:193–204

Rao AS (1968) Biology of *Polymyxa graminis* in relation to soil-borne wheat mosaic virus. Phytopathology 58:1516–1521

Rao AS, Brakke MK (1969) Relation of soil-borne wheat mosaic virus and its fungal vector, *Polymyxa graminis*. Phytopathology 59:581–587

Sahtiyanci S (1962) Studien über einige wurzelparasitäre Olpidiaceen. Arch Mikrobiol 41:187–228

Slogteren DHM van, Visscher HR (1967) Transmission of a tobacco necrosis virus, causing "Augusta disease" to the roots of tulip by zoospores of the fungus *Olpidium brassicae* (Wor.) Dang. Meded Rijksfac Landbouwwet Gent 32:927–938

Slykhuis JT (1976) Wheat spindle streak mosaic virus. CMI/AAB descriptions of plant viruses no 167, pp 3

Slykhuis JT, Barr DJS (1978) Confirmation of *Polymyxa graminis* as a vector of wheat spindle
    streak mosaic virus. Phytopathology 68:639–643
Smith PR, Campbell RN, Fry PR (1969) Root discharge and soil survival of viruses. Phytopathology
    59:1678–1687
Stocky G, Vuittenez A, Putz C (1977) Mise en évidence in situ de particules en batonnets, corre-
    spondant probablement au virus de la Rhizomanie, dans les tissus de diverses Chénopodiacées
    inoculées expérimentalement, ainsi que dans le champignon *Polymyxa betae* Kesk., myxomycete
    parasitant les racines de plantes malades. Ann Phytopathol 9:536–537 (abstr)
Tamada T (1975) Beet necrotic yellow vein virus. CMI/AAB descriptions of plant viruses no 144,
    pp 4
Tamada T, Abe H, Baba T (1975) Beet necrotic yellow vein virus and its relation to the fungus,
    Polymyxa betae. Proc 1st Intersect Congr Int Assoc Microbiol Soc 3:313–320
Taylor CE, Robertson WM (1977) Virus vector relationships and mechanics of transmission. Proc
    Am Phytopathol Soc 4:20–29
Teakle DS (1962) Transmission of tobacco necrosis virus by a fungus, *Olpidium brassicae*. Virology
    18:224–231
Teakle DS, Gold AH (1963) Further studies of *Olpidium* as a vector of tobacco necrosis virus.
    Virology 19:310–315
Teakle DS, Hiruki C (1964) Vector specificity in *Olpidium*. Virology 24:539–544
Temmink JHM (1971) An ultrastructural study of *Olpidium brassicae* and its transmission of to-
    bacco necrosis virus. Meded Landbouwhogesch Wageningen 71:1–135
Temmink, JHM, Campbell RN (1968) The ultrastructure of *Olpidium brassicae*. I. Formation of
    sporangia. Can J Bot 46:951–956
Temmink JHM, Campbell RN (1969a) The ultrastructure of *Olpidium brassicae*. II. Zoospores.
    Can J Bot 47:227–231
Temmink JHM, Campbell RN (1969b) The ultrastructure of *Olpidium brassicae*. III. Infection of
    host roots. Can J Bot 47:421–424
Temmink JHM, Campbell RN, Smith PR (1970) Specificity and site of in vitro acquisition of
    tobacco necrosis virus by zoospores of *Olpidium brassicae*. J Gen Virol 9:201–213
Thottappilly G (1972) Weitere Übertragungsversuche mit dem falschen Blattrollvirus der Erbse.
    Z Pflanzenkr Pflanzenschutz 79:10–14
Thottappilly G, Schmutterer H (1968) Zur Kenntnis eines mechanisch-, samen-, pilz- und insekten-
    übertragbaren neuen Virus der Erbse. Z Pflanzenkr Pflanzenschutz 75:1–8
Thouvenel JC, Dollet M, Fauquet C (1976) Some properties of peanut clump, a newly discovered
    virus. Ann Appl Biol 84:311–320
Tomlinson JA, Garrett RG (1962) Role of *Olpidium* in the transmission of big vein disease of
    lettuce. Nature (London) 194:249–250
Tomlinson JA, Garrett RG (1964) Studies on the lettuce big-vein virus and its vector *Olpidium
    brassicae* (Wor.) Dang. Ann Appl Biol 54:45–61
Tomlinson JA, Smith BR, Garrett RG (1962) Graft transmission of lettuce big vein. Nature (Lon-
    don) 193:599–600
Toyama A, Kusaba T (1970) Transmission of soil-borne barley yellow mosaic virus. 2. *Polymyxa
    graminis* Led. as vector. Ann Phytopathol Soc Jpn 36:223–229
Vuittenez A, Arnold J, Spindler C, de Larambergue H (1977) Test de transmission du virus de la
    rhizomanie de la betterave (Beet necrotic yellow vein) par le champignon *Polymyxa betae*
    (Kesk.) myxomycète associé aux racines. Application à la prospection de la maladie. Ann
    Phytopathol 9:537 (abstr)
Westerlund FV, Campbell RN, Grogan RG (1978a) Effect of temperature on transmission, transloca-
    tion, and persistence of the lettuce big-vein agent and big-vein symptom expression. Phytopath-
    ology 68:921–926
Westerlund FV, Campbell RN, Grogan RG, Duniway JM (1978b) Soil factors affecting the repro-
    duction and survival of *Olpidium brassicae* and its transmission of big vein agent to lettuce.
    Phytopathology 68:927–935
Yarwood CE (1971) Erysiphaceae transmit virus to *Chenopodium*. Plant Dis Rep 55:342–344
Yarwood CE, Hecht-Poinar E (1973) Viruses from rusts and mildews. Phytopathology 63:1111–1115

Fungal Viruses in Edible Fungi

R. USHIYAMA

Tottori Mycological Institute, Kokoge-Hirohata-211, Tottori/Japan

1 Introduction

Since the first demonstration of fungal viruses associated with the "die-back disease" of cultivated mushroom, *Agaricus bisporus* (Lange) Sing. (Hollings, 1962; Hollings et al., 1963), a number of viruslike particles (VLPs) have been found in various species of fungi. Accumulating data based on physicochemical, biological, and ultrastructural properties suggest that VLPs are truly viruses. However, VLPs have not been shown to be infectious as free particles. Recent research on fungal viruses and VLPs has been reviewed by Hollings (1978) and Lemke (1976, 1977). This short article will attempt to review mainly ultrastructural evidence for viruses and VLPs found in edible fungi, particularly cultivated mushrooms, *A. bisporus* and *Lentinus edodes,* with some emphasis on work from the author's own laboratory.

2 Presence of Viruses and Viruslike Particles in Hosts

Fungal viruses or VLPs have been found in several species of fungi, and their presence has not been always accompanied by abnormal growth and morphology of vegetative mycelia and basidiocarps, suggesting that they are predominantly latent in their hosts (Lemke and Nash, 1974; Wood, 1975; Lemke, 1976, 1977). In contrast, mushroom viruses of *Agaricus bisporus* cause disease symptoms showing typical abnormalities (Gandy, 1960; Hollings et al., 1963; Last et al., 1967; Schisler et al., 1967; Dieleman-van Zaayen and Temmink, 1968): (1) Mycelial isolates from virus-infected sporophores show a slow growth, brown velvety texture, and no coarse strands on malt agar. (2) Virus-infected sporophores exhibit dwarfing, early maturity, watery stipes, hard gills, and thickened, barrel-shaped stipes. These symptoms are highly variable, probably depending on environmental conditions.

Five viruses have been isolated from diseased mushrooms of *A. bisporus* in Britain (Hollings, 1972) and some of them have now been found in other countries where mushrooms are extensively cultivated (Table 1) (Schisler et al., 1967; Dieleman-van Zaayen and Temmink, 1968; Albouy, 1972; Saksena, 1975; Sinden, 1975; Yamashita et al., 1975; Moyer and Smith, 1977). Four of the viruses are polyhedral in shape and measure 25, 29, 34–35 and 50 nm diameter; one is a bacilliform particle 19 × 50 nm, similar in appearance to the long particles of alfalfa mosaic virus (AMV). However, the bacilliform particle is not serologically related to AMV (Hollings, 1962; Hollings, 1978). Hollings (1978) suggested that this virus might be a candidate for membership of the hitherto monotypic group containing AMV. Most of the available evidence suggests that at least three morphological types of particle (25, 34—35 and 19 × 50 nm) generally occur in this fungus singly or in combination. However, the relationships between the three types of particle and specific disease symptoms are not yet clearly established. The other form of spherical particles ca. 50 nm diameter seems to be relatively uncommon and usually occurs in very low concentration (Albouy, 1972; Hollings, 1972). Rodshaped particles 17 × 350 nm, morphologically similar to Tobamoviruses, have been reported from *A. bisporus* (Dieleman-van Zaayen, 1967; Albouy, 1972). Similar particles have been also observed in dip preparations and in sections of *Lentinus edodes* (Inoue et al., 1970; Ushiyama and Hashioka, 1973; Yamashita et al., 1975; Ushiyama and Nakai, 1975a) and other species of fungi (Dieleman-van Zaayen, 1967; Huttinga et al., 1975). From *Lentinus edodes,* an important species of edible fungi in Japan, spherical to polyhedral virus particles measuring ca. 25, 30, and 39 nm (Ushiyama and Nakai, 1975a) and ca. 30, 36, and 45 nm (Table 1; Yamashita et al., 1975) have been detected. However, there is no definitive evidence as to whether these reports refer to the same four viruses, or to five different viruses. These viruses occur in both healthy and abnormal sporophores (Ushiyama and Nakai, 1975a; Yamashita et al., 1975). There are data which suggest that mycelial abnormalities are correlated with a high virus concentration, particularly when *L. edodes* is grown on agar (Ushiyama and Nakai, 1975b; Ushiyama and Nakai, unpublished). We suggest that there appears to be some mechanism controlling the relationship between virus reproduction and hyphal growth.

Rod-shaped VLP's which vary in length and width, and exhibit a hollow central core filled with stain, have been observed in extracts and thin sections of *L. edodes* basidiocarps (Fig. 1) (Inoue et al., 1970; Ushiyama and Hashioka, 1973; Mori, 1975; Ushiyama, 1975; Yamashita et al., 1975).

The rods probably fall into two classes, rigid rods (25—28 × 280—310 nm) and long flexible rods (Table 1) (15—17 × 1500 nm in maximum length). The variation and discrepancy in particle length and width may be due to fragmentation of particles during isolation, the use of different negative stains, and various types of artifacts which occur during specimen preparation, as described by Matthews (1970). The rigid rods (25—28 × 280—310 nm), which were rarely found in *L. edodes* basidiocarps, occurred in insufficient concentration for physicochemical, serological, or ultrastructural examination. Therefore, it is not known whether or not they are viral in nature. The long flexible rods consist of a major polypeptide with a molecular weight of 23,000 daltons (Yamashita et al., 1975; Ushiyama, unpublished). When samples of tissues, taken from *L. edodes* basidiocarps containing spherical viruses (25, 30, and 39 nm) and rod-shaped particles (15 × 700—900 nm), were cultured for 15 days at 25°C on an agar medium, rodshaped particles were not observed in the vegetative mycelia (Ushiyama and Nakai, 1975a). The three types of spherical particles were first detected in mycelial tissue extracts taken of 4-day-old cultures. When mycelia were grown for more than 20 days, rod-shaped particles could be detected and increased in number with the age of fungus culture. These data raise the question of whether the long flexible rods are truly viruses.

No viruslike particles were found in other cultivated mushrooms in Japan, *Pholiota nameko, Pleurotus ostreatus, Flammulina velutipes, Auricularia auricula-judae,* and *Volvariella volvacea.*

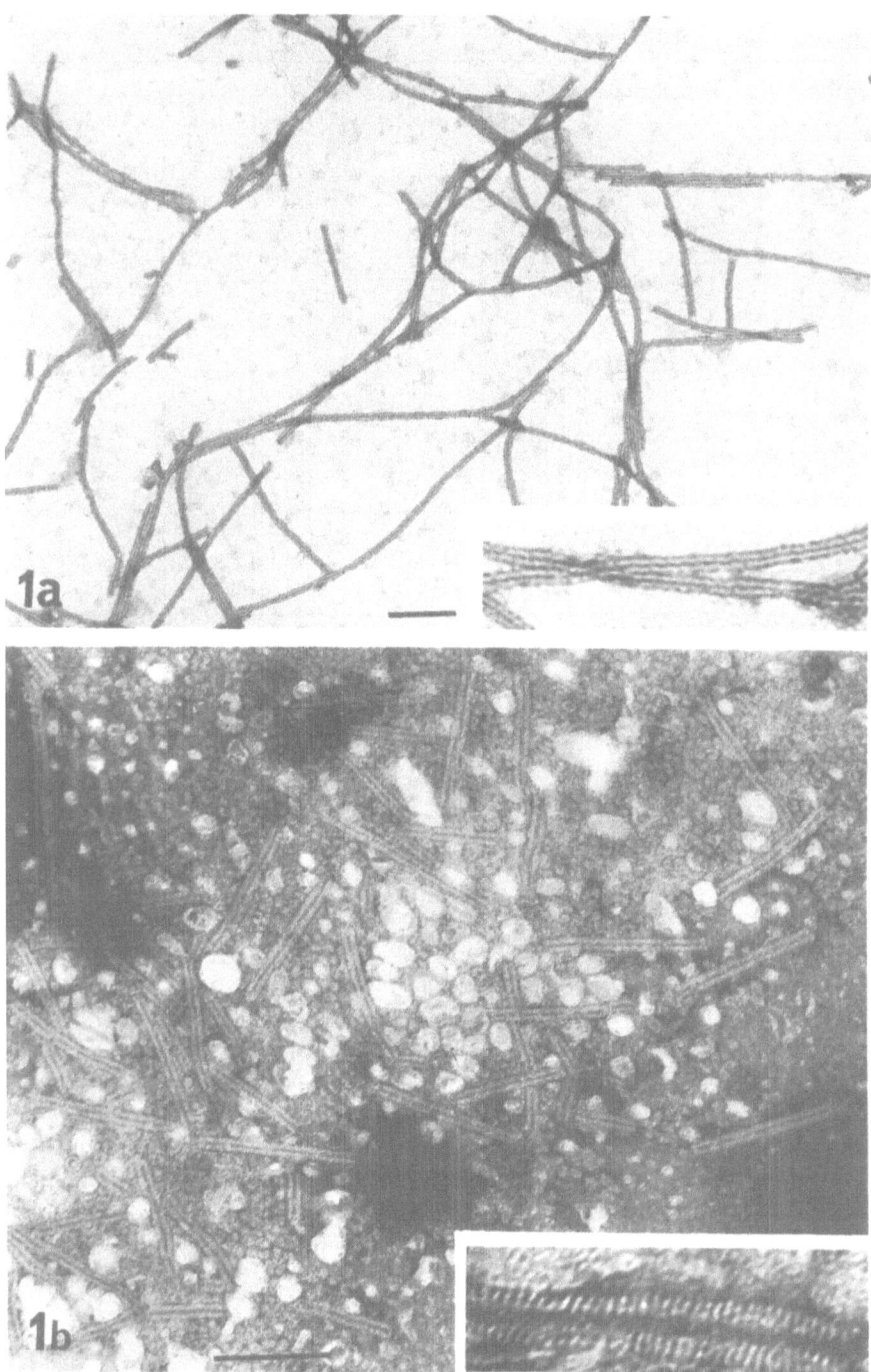

Fig. 1a, b. Electron micrographs showing rod-shaped viruslike particles in negatively stained preparations from *L. edodes.* **a** shows the long flexible rods possessing hollow central cores filled with stain. **b** exhibits the rigid rods with the helix. *Bar* = 200 nm

Table 1. Occurrence of viruses and viruslike particles in *A. bisporus* and *L. edodes*

Host species	Particle size (nm)	Key references[a]
A. bisporus	Spherical type	
	25 (dsRNA)	1, 2, 3, 4, 5, 6, 7, 9, 10, 11, 14
	29	1, 6
	34–35 (dsRNA)	1, 2, 3, 4, 7, 9, 10, 11
	50	1, 7
	Bacilliform type	
	19 × 50 (ssRNA)	3, 4, 6, 9, 10, 11, 12, 14
	Rod-shaped type	
	17 × 350	1, 5
L. edodes	Spherical type	
	25	13
	30	8, 13, 14
	36	8, 14, 15
	39 (dsRNA)	13
	45	14, 15
	Rod-shaped type	
	25–28 × 280–310	8, 13, 15
	15–17 × 100–1500	8, 13, 14, 15

[a] 1, Albouy (1972); 2, Atkey et al. (1975); 3, Dieleman-van Zaayen and Temmink (1968); 4, Dieleman-van Zaayen (1972); 5, Dieleman-van Zaayen (1967); 6, Hollings (1962); 7, Hollings (1972); 8, Inoue et al. (1970); 9, Marino et al. (1976); 10, Passmore and Frost (1974); 11, Moyer and Smith (1977); 12, Saksena (1975); 13, Ushiyama and Nakai (1975a); 14, Yamashita et al. (1975); 15, Mori (1975)

3 Double-Stranded RNA Viruses

The fungal viruses or VLPs so far examined in detail seem typically to contain double-stranded RNA (dsRNA) (Moffitt and Lister, 1973, 1975). The physicochemical evidence for the presence of dsRNA components has been presented for two of the spherical particles (25 and 34–35 nm) from *A. bisporus* (Molin and Lapierre, 1973; Barton, 1975; Barton and Hollings, in press) and for the 39-nm polyhedral particles from *L. edodes* (Fig. 2) (Ushiyama et al., 1976, 1977). In addition, serological evidence has been obtained for dsRNA in nucleic acids extracted from virus-infected mushrooms of *A. bisporus* (Marino et al., 1976). The bacilliform particles of 19 × 50 nm from *A. bisporus*, morphologically similar to AMV, contain single-stranded RNA (ssRNA) (Molin and Lapierre, 1973). Nonencapsidated ssRNA (free ssRNA) occurs in very high concentration in *L. edodes* (Ushiyama, unpublished). It would be of interest to know the origin of this free ssRNA.

4 Intracellular Nature and Behavior of Viruses in Hosts

The intracellular appearance and localization of viruses have been studied by electron microscopy of thin sections of *A. bisporus* (Dieleman-van Zaayen and Igesz, 1969; Albouy, 1972; Dieleman-van Zaayen, 1972) and *L. edodes* (Ushiyama and Nakai, 1975a, in press). Five morphological types of viruses (spherical 25, 29, 34–35, 50 nm, bacilliform 19 × 50 nm) from *A. bisporus* and 39-nm polyhedral virus from *L. edodes* have been observed. Detection of the other sizes of particles found in negatively stained preparations from *L. edodes* was difficult in thin sections, because their frequency of occurrence was extremely low. In addition, 25-nm particles were not distinguishable from cytoplasmic ribosomes in thin sections. There is now substantial evidence to show that: (1) No cytopathic effects on the intracellular organelles and cytoplasm are found in virus-infected *A. bisporus*. (2) In *A. bisporus* and *L. edodes,* virus particles are generally present in the ground cytoplasm, vesicles or vacuoles of host cells. The frequent occurrence of particle aggregates inside the cytoplasmic vesicles and vacuoles is a striking feature. (3) Cytoplasmic vesicles or vacuoles may be considered as sites of particle accumulation and not as the location of particle replication or assembly. (4) In *L. edodes,* the accumulated particles are crystallized possibly due to hyphal senescence (Fig. 3) (Ushiyama and Nakai, in press). Virus crystals appeared in sections of vegetative mycelium either in a square or hexagonal pattern, similar to that observed with higher plant viruses (Ushiyama, 1971; Russo et al., 1968). Fungal virus particles may be crystallized in situ essentially as a cubic close-packed structure, as described by Wyckoff (1948). (5) No virus particles are associated with nuclei, mitochondria, or other cellular organelles. (6) It is assumed that cell-to-cell movement of virus particles must take place via the dolipores which range ca. 100 nm to 130 nm in diameter in both basidiocarps

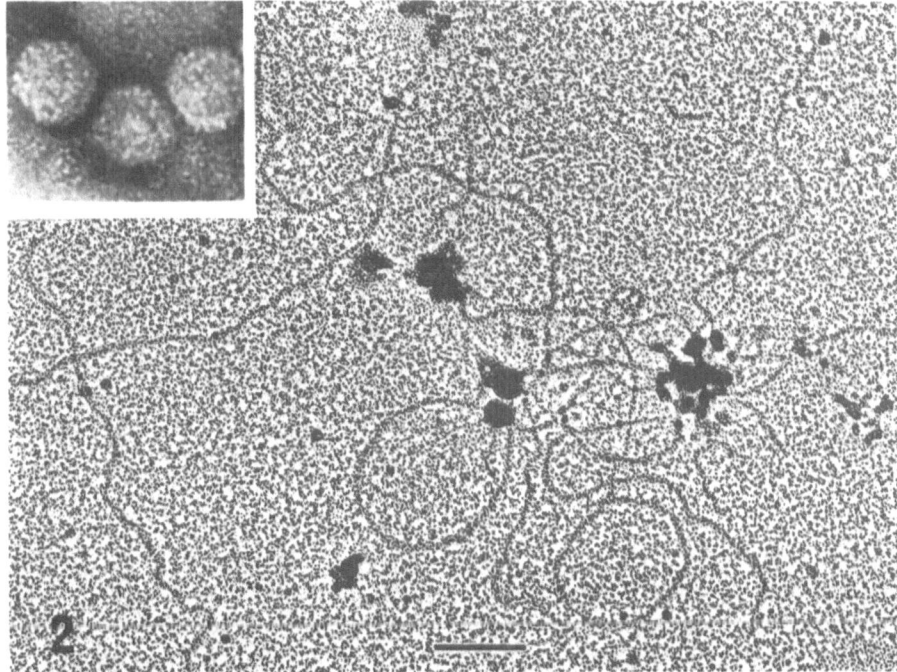

Fig. 2. Electron micrograph showing a spidery form of double-stranded RNA released from 39-nm particles *(inset)* in *Lentinus edodes.* (From Ushiyama et al., 1977). *Bar* = 150 nm

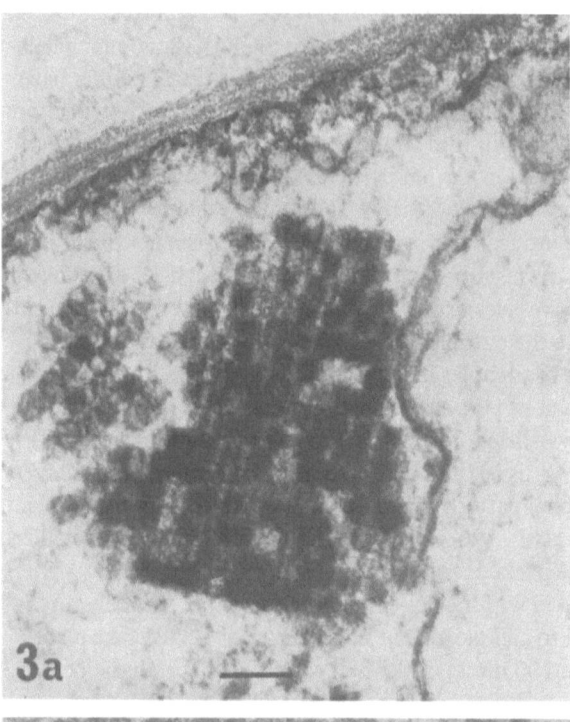

Fig. 3a, b. Crystals of 39-nm polyhedral virus particles in thin sections of *L. edodes* mycelium. *Bar* = 100 nm

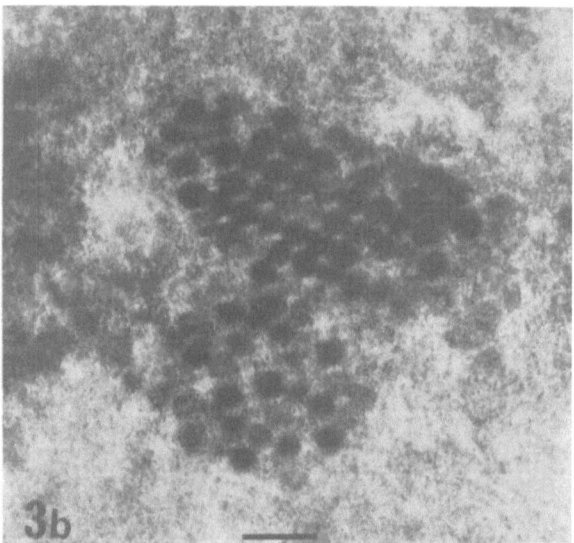

and vegetative mycelia of *A. bisporus* and *L. edodes*. These dimensions would be large enough to allow the passage of most viruses. The direct evidence for the movement of virus particles through dolipores has been demonstrated by Dieleman-van Zaayen (1972).

5 Concluding Remarks

Knowledge about the biological significance of fungal viruses in the hosts is much less advanced than for plant viruses. The main reason for this is the extreme difficulty of mechanical inoculation of cell-free virus particles into host cells. In plants, various disease symptoms must originate from cytological and biochemical aberrations induced directly or indirectly by the virus (Matthews, 1970). In *A. bisporus,* disease symptoms and loss in yield are induced by infection with viruses (Dieleman-van Zaayen and Temmink, 1968; Hollings and Stone, 1971; Last et al., 1967, 1974), but no cytological abnormalities have been definitely observed (Dieleman-van Zaayen and Igesz, 1969; Dieleman-van Zaayen, 1972). Most fungal viruses seem to be latent (Lemke, 1976, 1977; Hollings, 1978). Sometimes *L. edodes* mycelial samples exhibiting slow and degenerated growth contain more virus than healthy mycelia (Ushiyama and Nakai, 1975b). Therefore there may be a specific correlation between the amount of virus and abnormality of host cells.

The appearance and localization of mushroom viruses in *A. bisporus* and *L. edodes* have been studied by electron microscopy. The general conclusion from such observations is that virus particles are probably assembled in the ground cytoplasm. However, the cytopathological behavior of viruses within cells has not yet been established. It will be interesting to know how viruses cause disease symptoms in the growing sporophores and vegetative mycelia of *A. bisporus* and how virus replication is related to disease expression.

Acknowledgment. I wish to thank Dr. N. Hiratsuka of the Tottori Mycological Institute for his support of this work.

Summary

Five morphologically distinct virus particles have been found in diseased fruit bodies and vegetative mycelia of *Agaricus bisporus* (Lange) Sing. from several countries where mushrooms are extensively cultivated. Four of them are polyhedral particles 25, 29, 34–35, and 50 nm in diameter, one represents bacilliform particles 19 × 50 nm and is rather similar in appearance to the long particles of alfalfa mosaic virus. At least three types of 25, 34–35 and 19 × 50 nm particles often occur in combination, and disease symptoms are quite characteristic. However, the relationships between these particles and the disease are not clearly established. Recently two types of 25 and 34 nm particles have been shown to contain a double-stranded RNA (dsRNA). At present, the only fungal virus in cultivated mushrooms reported to be a single-stranded RNA is the bacilliform particle in this fungus.

On the other hand, four morphological types of polyhedral viruslike particles with diameters of 25, 30, 39, and 45 nm have been isolated in "Shiitake" mushrooms, *Lentinus edodes* (Berk.) Sing., which are commonly cultivated as an important species of edible fungi in Japan. However, it is uncertain whether or not these viruslike particles cause the deformed fruit bodies and degeneration symptoms of vegetative mycelium. Regarding biochemical properties of the viruslike particles in this fungus, the 39 nm particles possess a dsRNA which shows one band with high molecular weight in polyacrylamide gel electrophoresis.

Ultrastructural examinations of two species of edible fungi mentioned above have revealed that the virus particles (25, 34, and 19 × 50 nm in *A. bisporus* and 39 nm in *L. edodes*) are frequently distributed in the cytoplasm, in membrane-bound vesicles and in the vacuoles, but not in nuclei or mitochondria. The movement from cell to cell of virus particles most probably takes place through dolipores. The dimension of dolipores (ca. 120–300 nm in diameter) and parenthesome pores (ca. 80–100 nm in diameter) is large enough to allow the passage of most virus particles.

In other cultivated mushrooms, *Pholiota nameko*, *Flammulina velutipes* and *Pleurotus ostreatus*, viruslike particles have not been found until now.

References

Albouy J (1972) Étude ultramicroscopique du complexe viral de la "goutte sèche" de carpophores d'*Agaricus bisporus*. Ann Phytopathol 4 (1): 39–44

Atkey PT, Barton RJ, Hollings M, Stone OM (1975) Mushroom *Agaricus bisporus* (Lange) Sing. Rep. Glasshouse Crops Res Inst for 1974, pp 120–121

Barton RJ (1975) Purification and some properties of two viruses infecting *Agaricus bisporus:* the cultivated mushroom. 3rd Int Congr Virol (Madrid), Abstr p 147

Dieleman-van Zaayen A (1967) Virus-like particles in a weed mould growing on mushroom trays. Nature (London) 216: 595–596

Dieleman-van Zaayen A (1972) Intracellular appearance of mushroom virus in fruiting bodies and basidiospores of *Agaricus bisporus*. Virology 47: 94–104

Dieleman-van Zaayen A, Igesz O (1969) Intracellular appearance of mushroom virus. Virology 39: 147–157

Dieleman-van Zaayen A, Temmink JHM (1968) A virus disease of cultivated mushrooms in the Netherlands. Neth J Plant Pathol 74: 48–51

Gandy DG (1960) "Watery stipe" of cultivated mushrooms. Nature (London) 185: 482–483

Hollings M (1962) Viruses associated with a die-back disease of cultivated mushroom. Nature (London) 196:692–695

Hollings M (1972) Recent research on mushroom viruses. Mushroom Sci 8: 733–738

Hollings M (1978) Mycoviruses: Viruses that infect fungi. Adv Virus Res 22: 1–53

Hollings M, Stone OM (1971) Viruses that infect fungi. Annu Rev Phytopathol 9: 93–118

Hollings M, Gandy DG, Last FT (1963) A virus disease of fungus: die-back of cultivated mushroom. Endeavour 22: 112–117

Huttinga H, Wichers HJ, Dieleman-van Zaayen A (1975) Filamentous and polyhedral viruslike particles in *Boletus edulis*. Netherl J Plant Pathol 81: 102–106

Inoue T, Furatani H, Nishikado Y (1970) Virus-like particles in *Lentinus edodes*. Ann Phytopathol Soc Jpn 36: 356

Last FT, Hollings M, Stone OM (1967) Some effects of cultural treatments on virus diseases of cultivated mushroom *Agaricus bisporus*. Ann Appl Biol 59: 451–462

Last FT, Hollings M, Stone OM (1974) Effects of cultural conditions on mycelial growth of healthy and virus-infected cultivated mushroom, *Agaricus bisporus*. Ann Appl Biol 76: 99–111

Lemke PA (1976) Viruses of eukaryotic microorganisms. Annu Rev Microbiol 30: 105–145

Lemke PA (1977) Fungal viruses and agriculture. Virology in agriculture. Allanheld Osmun & Co, Montclair, New York, pp 159–175

Lemke PA, Nash CH (1974) Fungal viruses. Bacteriol Rev 38: 29–56

Marino R, Saksena KN, Schuler M, Mayfield JE, Lemke PA (1976) Double-stranded ribonucleic acid in *Agaricus bisporus*. Appl Environ Microbiol 31: 433–438

Matthews REF (1970) Plant virology. Academic Press, New York, p 778

Molin G, Lapierre H (1973) L'acide nucleique des virus de champignons: Cas de virus de l'*Agaricus bisporus*. Ann Phytopathol 5: 233–240

Moffitt EM, Lister RM (1973) Detection of mycoviruses using antiserum specific for dsRNA. Virology 52: 301–304

Moffitt EM, Lister RM (1975) Application of a serological screening test for detecting double-stranded RNA mycoviruses. Phytopathology 68: 851–859

Mori K (1975) Studies of virus-like particles in *Lentinus edodes*. Proc Ist Intersec Congr IAMS 3: 396–401

Moyer JW, Smith SH (1977) Purification and serological detection of mushroom virus-like particles. Phytopathology 67: 1207–1210

Passmore E, Frost RR (1974) The detection of virus-like particles in mushrooms and mushroom spawns. Phytopathol Z 80: 85–87

Russo MS, Martelli GP, Quacquarelli A (1968) Studies on the agent of artichoke mottle crinkle. IV. Intracellular localization of virus. Virology 34: 679–693

Saksena KN (1975) Isolation and large-scale purification of mushroom viruses. Dev Ind Microbiol 16: 134–144

Schisler LC, Sinden JW, Sigel EM (1967) Etiology, symptomatology, and epidemiology of a virus disease of cultivated mushrooms. Phytopathology 57: 519–526

Sinden JW (1975) Pathogenicity of mushroom viruses. Dev Ind Microbiol 16: 123–127

Ushiyama R (1971) Crystallization of turnip yellow mosaic virus in wilted leaves of chinese cabbage by treatment at high temperature. Japan J Microbiol 15 (3): 257–264

Ushiyama R (1975) Virus-like particles in shiitake mushroom, *Lentinus edodes* (Berk.) Sing. Proc Ist Intersec Congr IAMS 3: 402–406

Ushiyama R, Hashioka Y (1973) Viruses associated with Hymenomycetes. 1. Filamentous virus-like particles in the cells of a fruit body of shiitake, *Lentinus edodes* (Berk.) Sing. Rep Tottori Mycol Inst (Japan) 10: 797–805

Ushiyama R, Nakai Y (1975a) Viruses associated with Hymenomycetes. 2. Presence of polyhedral virus-like particles in shiitake mushroom, *Lentinus edodes* (Berk.) Sing. Rep Tottori Mycol Inst (Japan) 12: 53–60

Ushiyama R, Nakai Y (1975b) Viruses associated with Hymenomycetes. 3. Growth of *Lentinus edodes* mycelium associated with polyhedral virus-like particles. Ann Phytopathol Soc Jpn 41: 287

Ushiyama R, Nakai Y, Ikegami M (1976) Detection of double-stranded RNA from polyhedral virus-like particles in *Lentinus edodes* (Berk.) Sing. Proc Jpn Acad 52: 450–452

Ushiyama R, Nakai Y, Ikegami M (1977) Evidence for double-stranded RNA from polyhedral virus-like particles in *Lentinus edodes* (Berk.) Sing. Virology 77: 880–883

Yamashita S, Doi Y, Yora K (1975) Electron microscopic study of several fungal viruses. Proc Ist Intersec Congr IAMS 3: 340–350

Wood HA (1975) Isolation and characterization of viruses from fungi. Proc Ist Intersec Congr IAMS 3: 362–379

Wyckoff RWG (1948) The electron micrography of macromolecular crystals. Acta Crystallogr 1: 292–294

Infectivity and Transmission of Fungal Viruses

H. LECOQ[1] , M. BOISSONNET-MENÈS[2] , and P. DELHOTAL[3]

[1] INRA. Station de Pathologie Végétale, 84140 Montfavet/France. [2] Laboratoire de Cryptogamie, Université de Paris-Sud, Bât. 402, 91405 Orsay/France. [3] INRA. Station de Pathologie Végétale, 78000 Versailles/France

1 Introduction

The discovery of viruses infecting Macromycetes by Hollings, in 1962, and later, of viruses infecting Micromycetes by Ellis and Kleinschmidt, in 1967, has opened a new field of research for many laboratories around the world. Indeed, the importance of research on mycoviruses — i.e., viruses that infect fungi — presents many different aspects: (1) mycoviruses are pathogens of the cultivated mushroom, *Agaricus bisporus*, and they have been associated with a disease called "La France disease", responsible for serious yield losses; (2) mycoviruses are potential pathogens of fungi such as *Penicillium* sp. and *Aspergillus* sp. used by the microbiological industry and they may affect antibiotic production; (3) mycoviruses have been found in phytopathogenic fungi such as *Pyricularia oryzae, Colletotrichum lindemuthianum,* and *Gaeumannomyces graminis,* and they have been studied because of their possible use as biological control agents; (4) some of the dsRNA-containing mycoviruses are found in large amounts in their host cells, and therefore might be used for industrial production of dsRNA, which is an important interferon inducer in mammals (for a general review on mycoviruses see: Bozarth, 1972; Lemke and Nash, 1974; Hollings, 1978).

Viruses have been found now in over 100 fungal species, but in most instances, they have been only observed by electron microscopy, without further characterization of the viral particles. In some cases, however, accurate biochemical and biophysical data have been obtained for the isometric dsRNA viruses. Little is known about the biological properties of these viruses, including their effect on host metabolism, their means of transmission, and their host range. This is why searchers have felt it prudent to call these nucleoproteins "viruslike particles" (VLP's), all the conditions of Koch's postulate not being fulfilled in most cases. Nevertheless, for terminological simplification, we will refer to them as viruses.

We will discuss here the major aspects of mycovirus transmission, with special attention to the epidemiology and infectivity of these viruses.

2 Mycovirus Transmission Through Spores

2.1 Transmission Through Conidiospores

Transmission through conidiospores seems to be a common feature for the viruses infecting Micromycetes (Table 1). Unfortunately we lack definite evidence for many other viruses , in which this way of transmission appears likely, but remains unproven.

Table 1. Micromycetes which transmit viruses through conidiospores

Aspergillus flavus	Wood et al. (1974)
A. foetidus	Chang and Tuveson (1973)
Colletotrichum lindemuthianum	Delhotal et al. (1976)
Fusarium roseum culmorum	Lapierre (unpublished)
Gaeumannomyces graminis	Rawlinson et al. (1977)
Penicillium brevicompactum	Sansing et al. (1973)
P. chrysogenum	Lemke et al. (1973)
P. claviforme	Metitiri and Zachariah (1972)
P. cyaneo-fulvum	Banks et al. (1969)
P. stoloniferum	Banks et al. (1968)
Pyricularia grisea	Lecoq (unpublished)
P. oryzae	Ferault et al. (1971)

Virus particles have been observed in ultrathin sections of conidiospores of *Penicillium stoloniferum* and *P. brevicompactum* (Hooper et al., 1972) but not in sections of conidiospores from *P. chrysogenum,* this apparently being due to the dense cytoplasm of the spores (Yamashita et al., 1973). Viruses similar to those found in mycelium have been extracted from conidiospores of *P. stoloniferum, P. brevicompactum* and *P. chrysogenum* (Sansing et al., 1973). In addition, no major differences were found between the amount of dsRNA extracted from spores and mycelium of these species, as compared to that extracted from the respective mycelium, indicating that similar amounts of virus are found in conidiospores and in mycelia.

The rate of virus transmission through conidiospores seems to be very high. In *P. stoloniferum,* Bozarth et al. (1971) reported that 12 out of 12 monoconidial isolates from an infected culture, were infected, and in *P. chrysogenum* an average of 9 monoconidial isolates out of 10 were infected (Crosse and Mason, 1974). In *Pyricularia oryzae* 100 monoconidial isolates out of 100 from an infected strain were found to be virus-infected (Boissonnet-Menès and Lecoq, 1976). More recently, De Marini et al. (1977) have studied extensively the transmission of *P. stoloniferum* virus slow (PsV-S) and *P. stoloniferum* virus fast (PsV-F) through conidiospores of an isolate of *P. stoloniferum.* It appeared that 5% of the 43 monoconidial isolates studied were virus-free, 2% contained only the PsV-S virus, and 93% of the isolates contained both viruses, but with various amounts of PsV-F. When the inoculum consisted of fragments of mycelium or masses of spores, the virus concentration in the cultures stayed rather constant. These results could be explained by an uneven distribution of viruses in the mycelium, and/ or, by a delay in virus translocation from hyphae to conidiophores and then to conidiospores. A conidiospore could thus be released before any infective viral particle had reached its protoplasm. Conidiospore size, as well as differences in conidiogenesis mechanisms, may also induce variation of the virus transmission rates in different fungal species.

In addition to these results, viruses have been observed in structures of vegetative multiplication including chlamydospores (*Mycogone perniciosa,* Albouy and Lapierre, 1972), uredospores (*Puccinia graminis* f. sp. *tritici,* Rawlinson and Maclean, 1973, and *Uromyces phaseoli* f. sp. *vignae,* MacDonald and Heath, 1978) and sclerotia (*Sclerotium cepivorum,* Lapierre et al., 1971). Irregular transmission of dsRNA through conidiospores has also been reported in *Endothia parasitica* (Day et al., 1977).

2.2 Transmission Through Basidiospores

Virus transmission through basidiospores was shown in *Agaricus bisporus* by Schisler
et al. (1963), and mushroom virus 4 has been observed in ultrathin sections of basidiospores by Dieleman-van Zaayen (1972). No data are available for the rate of virus transmission through basidiospores of *A. bisporus.* However, 10 spores from an infected car-
pophore were enough to initiate the disease in experimental mushroom culture beds
(Schisler et al., 1967). In *Lentinus edodes,* the 25, 30, and 39 nm particles are transmitted through basidiospores (Ushiyama and Nakai, 1975) and a 130 nm virus was similarly transmitted in *Coprinus lagopus* (Shahriari et al., 1973).

2.3 Transmission Through Ascospores

Of particular interest was the report by Lapierre et al. (1970), that viruses were not
transmitted through ascospores of *Gaeumannomyces (Ophiobolus) graminis.* This finding was confirmed by Rawlinson et al. (1973). They observed that 56 single ascospore
cultures from 4 virus-containing strains were devoid of viral particles. Recently Férault
and Tivoli (personal communication) have found that 7 out of 8 ascospores produced
by an infected strain gave rise to infected cultures. The discrepancy in these results may
be interpreted either by genetic differences between the strains (some of them bearing
a "transmission" gene and the others not), by differences in virus strains or in virus concentration in the hosts.

2.4 Significance of Mycovirus Transmission Through Spores

The presence of viruses in spores does not seem to decrease their viability. Indeed, conidiospores produced by an experimentally infected strain were shown to be as viable as
those produced by the same but healthy strain of *Colletotrichum lindemuthianum* (Delhotal et al., 1976). No difference was found in the viability of spores produced by
healthy and infected strains of *P. stoloniferum* (De Marini et al., 1977). In *Agaricus
bisporus,* basidiospores produced by infected carpophores germinate faster and more
abundantly than those produced by healthy carpophores, (Schisler et al., 1967; Dieleman-van Zaayen, 1970). They are still viable and able to initiate the disease after 5 1/2
years at 4°C or 3 1/2 years at laboratory temperature (Dieleman-van Zaayen, 1974;
Atkey et al., 1975). Because of the generally very high proportion of spores giving rise
to infected thalli, virus transmission through spores appears to be of major importance
for the maintenance of mycoviruses in nature. On the other hand, if virus infection
leads to a selective disadvantage for the fungal host, even when the rate of virus transmission through spores is 100%, virus-infected strains would be eliminated in the absence of other means of virus transmission such as by anastomoses or vectors.

3 Mycovirus Transmission Through Plasmogamy

Plasmogamy is achieved in fungi through anastomoses between hyphae produced by
different strains of a fungus. Plasmogamy may, or may not, be followed by the forma-

tion of heterokaryotic lines and/or by karyogamy. Electron microscopic examinations have shown that mycoviruses are found generally in large amounts in the hyphal cytoplasm (Albouy and Lapierre, 1971; Dieleman-van Zaayen, 1972; Hooper et al., 1972; Yamashita et al., 1973). Therefore it appears likely that mycoviruses may be readily transmitted from an infected strain to a healthy compatible one through fusion of cytoplasm. Often anastomoses do not occur in vitro because of genetic incompatibility, or of unsuitable medium conditions. Such is the case for *Pyricularia oryzae* in which heteroanastomoses have been rarely observed (Giatcong and Frederiksen, 1969). For such cases, plasmogamy may be achieved in vitro through protoplast fusions.

3.1 Anastomoses as a Means of Virus Transmission in Agaricus bisporus

Gandy (1960) and Gandy and Hollings (1962) obtained the first evidence for transmission by hyphal anastomoses of the viruses associated with "La France disease" in cultivated mushrooms. When a diseased slow-growing culture and a healthy one were allowed to develop side by side in the same Petri-dish, anastomoses were observed between hyphae from the healthy and the infected cultures. Subsequent transfers of mycelium from the initially healthy culture gave rise to slow-growing colonies from which viruses could be recovered. Later Schisler et al. (1967) demonstrated that anastomoses between healthy mycelium and germinating infected spores were frequent, particularly when the very young tip of a new germ tube comes into contact with the growing tip of the hyphae, regardless of the type of strain confronted (cream or white). However, transmission of the disease through anastomoses between mycelia of cream and white varieties seems to be rare (Kneebone et al., 1962).

Therefore virus spread may be achieved in *A. bisporus* either through infected mycelia which remain in trays after a crop and which would then anastomose with the healthy mycelium of the following crop (assuming that the strains are compatible) or by spores. If infected spores released by diseased carpophores (which, incidentally, open earlier and shed spores quicker than the healthy ones) are introduced into the mushroom beds between spawning and casing (the most sensitive stage of the culture), they will initiate the disease which will develop radially from the inoculation point (Schisler et al., 1967; Dieleman-van Zaayen, 1970). Strict sanitary measures have been proposed to limit virus transmission in mushroom cultures, including disinfection of the culture trays, baskets, corridors, and the use of spore filters (Dieleman-van Zaayen, 1970). Recently the use of *A. bitorquis,* immune to mushroom virus 1, 3, and 4, has been proposed (Van Zaayen, 1976). Virus transmission through anastomoses has also been reported in another Agaricale, *Schizophyllum commune* (Koltin et al., 1973).

3.2 Heterokaryosis as a Means of Virus Transmission in Micromycetes

Anastomosis followed by heterokaryosis has been used to transmit mycoviruses in several Micromycetes. Before such experiments, however, it is necessary to have a virus-free strain (checked with the more accurate techniques for detection of viruses at successive time intervals) and markers which allow easy recognition of the initially healthy strain from the infected one. In most instances no clear morphological characteristic is associated with virus infection of fungi (contrary to the slow-growing aspect of infected cultures of *A. bisporus*). The markers generally used were either morphological differences, such as color, different trophic requirements, resistance to drugs, mating types, or a coloration by a fluorescent dye to differentiate healthy from infected hyphae (see Table 2). After anastomosis, and, eventually heterokaryosis, the strain bearing the phe-

Table 2. Micromycetes which transmit viruses through anastomoses

Fungus	Markers used	Reference
Aspergillus niger	Auxotrophy	Lhoas (1970)
Colletotrichum lindemuthianum	Auxotrophy, fungicide resistance	Delhotal et al. (1976)
Endothia parasitica [a,b]	Morphology	Day et al. (1977)
Gaeumannomyces graminis [b]	Fluorescent dye on infected strain	Rawlinson et al. (1973)
Penicillium chrysogenum	Morphology, auxotrophy	Nash et al., in Lemke and Nash (1974)
P. claviforme [b]	Morphology	Metitiri and Zachariah (1972)
P. stoloniferum [c]	Morphology	Lhoas (1971a)
Ustilago maydis	Mating types	Wood and Bozarth (1973)

a Only dsRNA transmission was shown in this fungus
b Heterokaryosis was not proved in these cases
c Morgan and Oakley (1976) using the same strains were unable to observe heterokaryosis and to repeat these transmission experiments

notype of the initially healthy isolate must be reisolated and checked for the presence or absence of virus particles. The steps followed in *Colletotrichum lindemuthianum* virus 1 transmission experiments, are shown Figure 1 (Delhotal et al., 1976). Virus particles extracted from the initially healthy strain at the end of the experiment were shown to be identical to those found in the initially infected strain (in regard to size, serological properties, sedimentation coefficients in sucrose gradients, densities in Cs_2SO_4, electrophoretic mobilities, and dsRNA composition; Lecoq and Delhotal, 1976). The viruses were still present in the newly infected strain one year after the transmission experiments were achieved, and in an amount similar to that found in the initially infected strain, showing that viral multiplication occurred in this new host.

3.3 Virus Transmission Through Protoplast Fusion

In *Pyricularia oryzae* anastomoses between hyphae of different strains are rarely observed (Giatcong and Frederiksen, 1969) although they have been reported to occur between monoconidial isolates derived from the same culture (Genovesi and Magill, 1976; Fatemi and Nelson, 1977). The strains used in our study (Boissonnet-Menès and Lecoq, 1976) never showed anastomoses, even when hyphal fragments were allowed to grow side by side. Mixtures of mycelia or spores of different auxotrophic mutants

Days

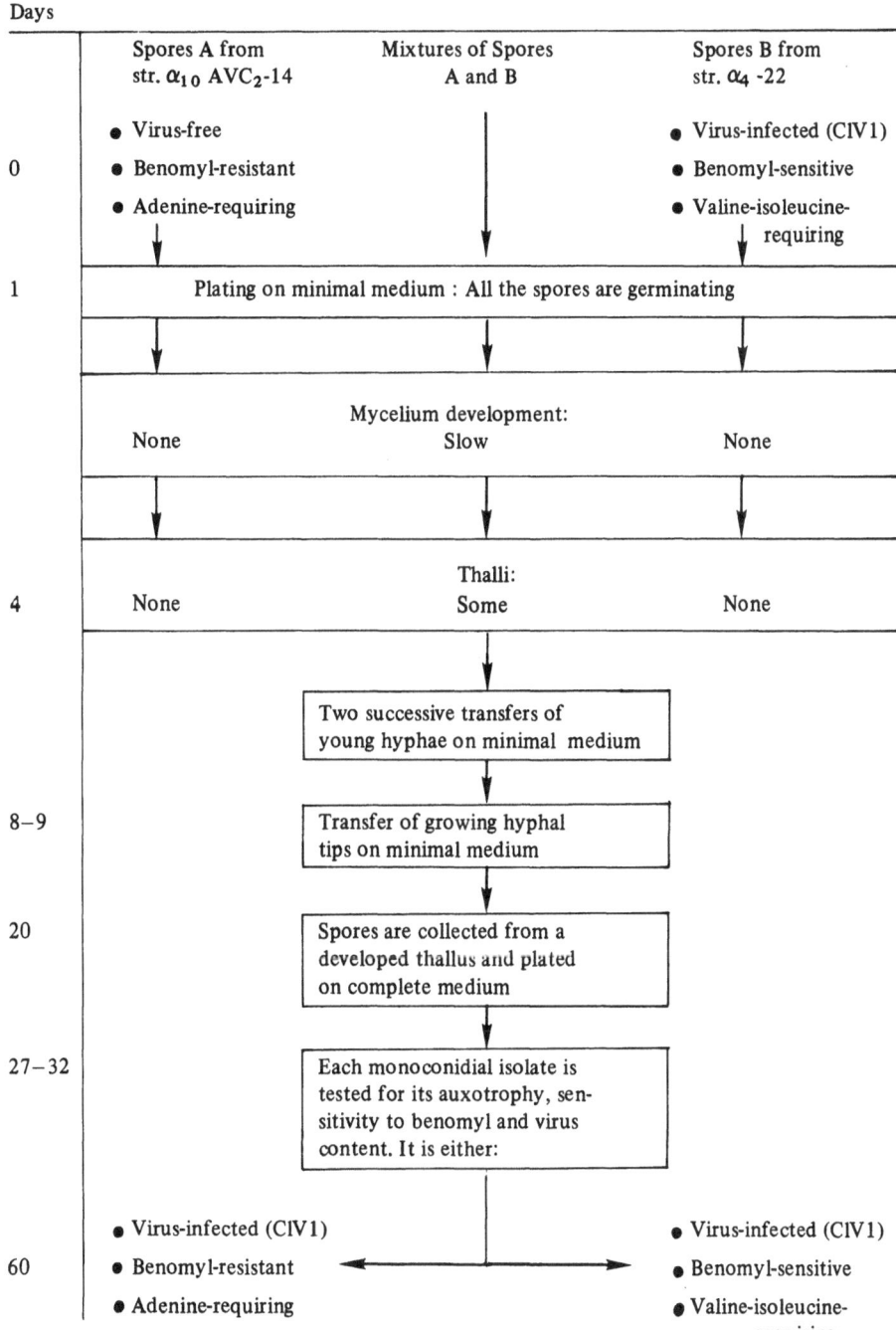

Fig. 1. Transmission of *Colletotrichum lindemuthianum* virus 1 through heterokaryosis. (Delhotal et al., 1976)

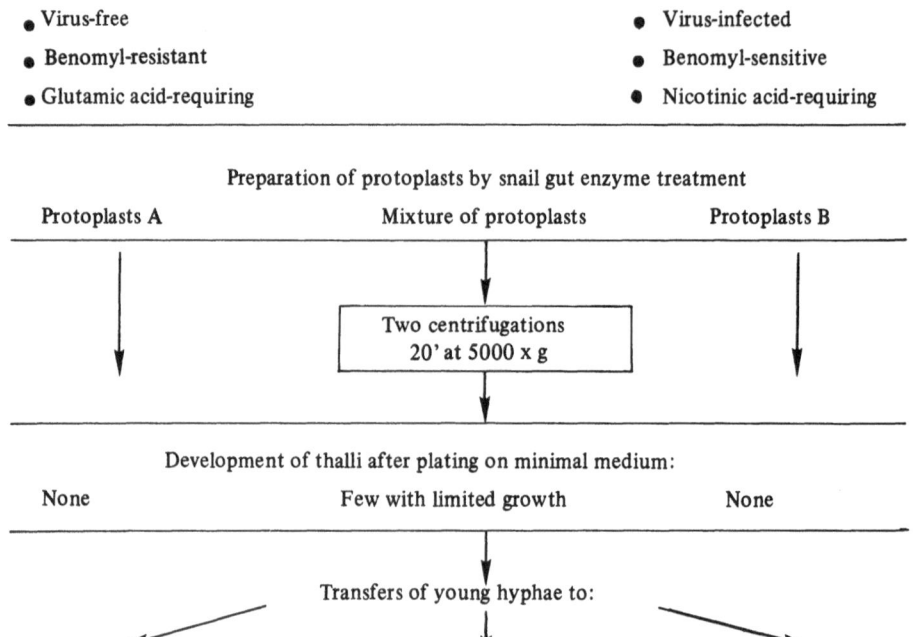

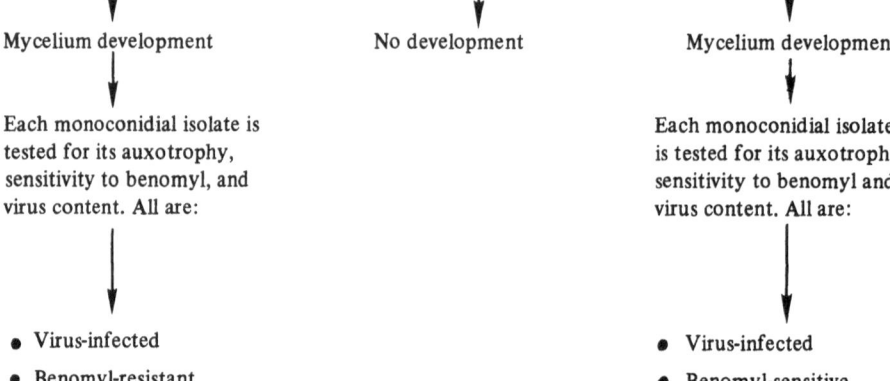

Fig. 2. Virus transmission through protoplast fusions in *Pyricularia oryzae*. (Boissonnet-Menès and Lecoq, 1976)

never gave rise to prototrophic thalli, indicating that heterokaryons were very rare, and/ or very unstable under these experimental conditions. Virus transmission through plasmogamy has nevertheless been obtained by inducing fusion of protoplasts from a virus-infected and a virus-free strain of *P. oryzae*. Details on the procedure used are given in Figure 2. The examination of extracts from the initially virus-free strain following protoplast fusion has shown the presence of viral particles similar in morphologies, serological properties, and in their sedimentation coefficients to those found in the initially infected strain.

Examinations made two years after these experiments have shown that viral particles were still present in the newly infected strains. Improvement in the rate of protoplast fusions may be made particularly by the use of polyethylene glycol (PEG) which has been reported to increase the number of protoplast fusions in various fungi (Ferenczy et al., 1976).

3.4 Significance of Virus Transmission Through Plasmogamy

Virus transmission through anastomoses has been shown to be of major importance in the epidemiology of the viral disease of *Agaricus bisporus*. Anastomoses could also be an important factor in the spread of other mycoviruses, particularly those found in phytopathogenic fungi. A demonstration of this could be obtained, for instance, by inoculating a susceptible host with a mixture of virus-infected and virus-free strains bearing different genetic markers, and then, after the incubation period, isolates of the originally virus-free type could be screened for the presence or absence of virus. However, Lemke and Nash (1974) pointed out that incompatibility between fungal strains may limit the transmission of mycoviruses under natural conditions, and, in this way, reduce their host range. Protoplast fusions may be a useful technique to increase this host range in vitro. Transmission experiments through plasmogamy have shown that fungal viruses are infectious, in the sense that they are able to invade a host previously devoid of them. Nevertheless these results do not indicate which is the infectious component: the nucleoproteins themselves, or another virus-associated component, which could also be transmitted through fusion of cytoplasms.

4 Mycovirus Transmission Through Vectors

Hollings (1967) reported a low level of mushroom virus 1 transmission by the phorid fly *Megaselia halterata* (Diptera), when aseptically reared flies were allowed to feed on purified virus and then transferred onto healthy mushroom mycelium. Nevertheless, no evidence has been reported for the transmission of virus by this fly from infected to healthy mycelia. Trials to transmit mushroom viruses using aseptically reared mites (*Tarsonemus myceliophagus*) (Hollings and Gurney, 1973), and to transmit *Gaeumannomyces graminis* viruses by mycophagous nematodes (Rawlinson and Muthyalu, 1974) have failed. Viruses have been found in fungi parasitic on other fungi including *Mycogone perniciosa* (Lapierre et al., 1972), *Verticillium fungicola* (Lapierre et al., 1973) and *Gonatobotrys* sp. (Spire et al., 1972), but here again there has been no evidence for the possible role of these fungi as mycovirus vectors. However, mycovirus transmission through vectors remains a possibility, and careful examination is needed to confirm or rule out this hypothesis, mainly if there is a high specificity of the vectors for the viruses they may transmit. On the other hand, mycoviruses may not need vectors, as they are usually easily transmitted through fungal spores, and able to infect healthy hosts through anastomoses.

5 Mycovirus Transmission Using Cell-Free Virus Preparations

5.1 Preliminary Attempts

Attempts to inoculate healthy mycelium with purified viruses by mechanical inoculation have not been successful. In *Pyricularia oryzae*, for instance, transmission was not obtained by mixing and shaking disrupted mycelium fragments, carborundum, and purified virus preparations. Lapierre and Faivre-Amiot (1970) and Hollings and Stone (1971) reported the same failure in *Sclerotium cepivorum* and in *Penicillium stoloniferum*, respectively. It appears that there are no specific adsorption sites on the hyphal cells, allowing for the penetration of virus particles.

A low level of virus transmission was, however, obtained when purified mushroom virus preparations were injected with a hypodermic syringe into healthy young *Agaricus bisporus* sporophores (Hollings, 1962; Dieleman-van Zaayen and Temmink, 1968). This technique would be difficult to apply to Micromycetes.

Lhoas (1972) reported success in infecting *Saccharomyces cerevisiae* by adding purified *Aspergillus niger* viruses or *P. stoloniferum* viruses to cells during the mating process. However the electrophoretic mobility of the virus particles extracted from yeast (after the *A. niger* virus transmission experiments) was different from those of the two viruses purified from *A. niger*. It is not known if the *A. niger* viruses were modified when they multiplied in the new host or if the viruses extracted from yeast were really yeast viruses which were present in low titre in one of the parental yeast strains. More data are needed to answer these questions.

5.2 Transmission Experiments Using Protoplasts

One of the main obstacles to mycovirus transmission seems to be the hyphal cell wall. The use of fungal protoplasts by removing this barrier would obviously be of great interest in infection experiments with purified viruses. Success in infecting healthy fungal protoplasts with purified viruses would (1) give a definitive proof that these nucleoproteins are infectious, (2) help to determine the effects which mycoviruses have on their hosts, (3) facilitate the study of the host range of these viruses and (4) by achieving synchronous infections, give the opportunity to elucidate the replicative cycle of these dsRNA viruses.

Lhoas (1971b) reported the transmission of PsV-S to virus-free *P. stoloniferum* protoplasts, using a partially purified virus preparation. Under these conditions, PsV-F was not transmitted. The amount of virus in the newly infected strain was about 10% of what was found in the initially infected isolate. This reduction in titre was already observed by the same author (1971a) in his heterokaryosis transmission experiments. Pallett (1972) has confirmed these results and extended them by infecting *P. chrysogenum*, *Marasmius androsaceus*, *Mucor hiemalis*, with *P. chrysogenum* viruses. However, in all the newly infected cultures, the virus titre was less than 1% of that found in the initially infected strain, and the level of virus fell below the limit of detectability after 3 years of subculturing (Pallett, 1976).

Besides these results there have been reported failures to transmit viruses to protoplasts. Bozarth (1975) failed to transmit viruses from *P. stoloniferum*, *P. chrysogenum*, and *Helminthosporium maydis* to protoplasts of healthy strains of the same species. Lai and Zachariah (1975) were unable to infect healthy protoplasts of *P. claviforme* with purified viral preparations, and Boissonnet-Menès and Lecoq (1976) also failed to infect protoplasts of the virus-free *Pyricularia oryzae* strain 1888 with *P. oryzae* viruses extracted from strain 95 P9. This virus-free strain was found to be able to support virus multiplication (see Sect. 2.3) and the purified virus preparations used in these experiments contained a large proportion of apparently intact particles (by electron microscopic examination and sucrose gradient analysis) and showed a high level of polymerase activity.

6 Concluding Remarks

The data available on fungal virus transmission using protoplasts are far too incomplete to make definitive interpretations. Many major questions remain unsolved.

Are purified nucleoproteins infectious? The results of Lhoas (1971b) and Pallett (1972) seem to indicate it; but it is not feasible to rule out the possibilities that a soluble component, necessary for infection, may have contaminated their partially purified virus preparations. On the other hand, the procedures necessary to obtain highly purified virus preparations may degrade particles, rendering them noninfectious.

Are all the particles of dsRNA multicomponent viruses necessary for infection? A particularly interesting model for determining this point would be the PsV-S in which particles with ssRNA, with dsRNA, and with ssRNA plus dsRNA have been found (Buck and Kempson-Jones, 1973). If the answer was no, then noninfective particles could bind to receptor sites on the protoplast membrane, and prevent or limit the penetration of infective particles.

Are fungal protoplasts always susceptible to infection? It is known that the susceptibility of plant protoplasts to a viral infection varies with the age of the plant and the position of the leaf from which they are prepared (Takebe, 1975). A similar situation could be true for fungal protoplasts and an optimal stage for protoplast production and infection should be looked for.

At present no answers are available for these questions. One of the main difficulties in transmission experiments is screening for infected thalli, as there are no clear-cut symptoms developed for infected mycelia in Micromycetes. The use of new sensitive and simple techniques to check for the presence or absence of viruses in small amounts of crude sap would allow for rapid screening for infected cultures. It would then be easy to examine a large number of samples and detect very low rates of virus transmission and very low titres of viruses in the newly infected cultures. For this purpose the ELISA technique (Clark and Adams, 1977) and the Immuno-Electron Microscopy (Milne and Luisoni, 1977) may be useful for research on fungal virus transmission.

Summary

Accurate biophysical and biochemical data have now been obtained for many viruses found infecting fungi but information concerning their means of transmission is rather scarce.

Mycoviruses have been shown to be transmitted through conidiospores in various Micromycetes, and in some cases with a very high transmission rate. Transmission also occurs through basidiospores, ascospores, and different vegetative multiplication organs.

Virus transmission through anastomoses appears to be of major importance for *Agaricus bisporus* virus spread in vivo. Experiments achieved in vitro have shown that viruses infecting Micromycetes are also transmitted through anastomoses. Incompatibility reactions between fungal strains may prevent anastomoses formation and then limit virus host ranges. Induced fusions of protoplasts from an infected and a healthy strain, a technique developed with *Pyricularia oryzae,* may be useful to transmit viruses when anastomoses between strains are not or only rarely observed.

Potential vectors such as insects, mites, nematodes, and hyperparasitic fungi have been investigated but the transmission of microviruses through vectors has not yet been proven.

Inoculation of healthy strains using partially purified virus preparations has been reported to lead to infection in *A. bisporus.* In Micromycetes, conflicting results have been reported, and there are, at the present time, no appropriate experimental conditions available to achieve reproducible infections of fungal protoplasts.

Nevertheless the successes met in transmitting mycoviruses lead to the conclusion that these viruses are infectious (i.e., they are able to multiply in a new and initially healthy host). In addition, virus transmission through spores, vectors and anastomoses are potential means of mycovirus spread in natural conditions.

References

Albouy J, Lapierre H (1971) Quelques aspects de l'infection virale chez les champignons (*Sclerotium cepivorum, Ophiobolus graminis, Agaricus bisporus*). Ann Univ A R E R S Reims 9: 333–339

Albouy J, Lapierre H (1972) Observation en microscopie electronique de souches virosées de *Mycogone perniciosa* agent d'une môle du champignon de couche. Ann Phytopathol 4:353–358

Atkey PT, Barton RJ, Hollings M, Stone OM (1975) Mushroom virus purification and characterization. In: Rep Glasshouse Crops Res Inst 1974, p 122

Banks GT, Buck KW, Chain EB, Himmelweit F, Marks JE, Tyler JM, Hollings M, Last FT, Stone OM (1968) Viruses in fungi and interferon stimulation. Nature (London) 218: 542–545

Banks GT, Buck KW, Chain EB, Darbyshire JE, Himmelweit F (1969) *Penicillium cyaneofulvum* and interferon stimulation. Nature (London) 222:89–90

Boissonnet-Menès M, Lecoq H (1976) Transmission de virus par fusion de protoplastes chez *Pyricularia oryzae.* Physiol Veg 14: 251–257

Bozarth RF (1972) Mycoviruses: a new dimension in microbiology. In: Environmental health perspectives. US Dep of Health, Education, and Welfare, Washington DC, pp 23–39

Bozarth RF (1975) The problem and importance of transmission of mycoviruses using cell free extracts. Proc Int Congr Virol (Madrid) Abstract, p.148

Bozarth RF, Wood HA, Mandelbrot A (1971) The *Penicillium stoloniferum* virus complex: two similar ds-RNA virus-like particles in a single cell. Virology 45: 516–523

Buck KW, Kempson-Jones GF (1973) Biophysical properties of *Penicillium stoloniferum* virus S. J Gen Virol 18: 223–235

Chang LT, Tuveson RW (1973) Genetics of strains of *Aspergillus foetidus* with virus-like particles. Genetics 74: s43–s44

Clark MF, Adams AN (1977) Characteristics of the microplate method of enzyme-linked immunosorbent assay for the detection of plant viruses. J Gen Virol 34: 457–483

Crosse R, Mason PJ (1974) Virus-like particles in *Penicillium chrysogenum.* Trans Br Mycol Soc 62: 603–610

Day PR, Dodds JA, Elliston JE, Jaynes RA, Anagnostakis SL (1977) Double-stranded RNA in *Endothia parasitica.* Phytopathology 67: 1393–1396

Delhotal P, Legrand-Pernot F, Lecoq H (1976) Etude des virus de *Colletotrichum lindemuthianum:* II Transmission des particules virales. Ann Phytopathol 8: 437–448

De Marini DM, Kurtzman CP, Fennell DI, Worden KA, Detroy RW (1977) Transmission of PsV-F and PsV-S mycoviruses during conidiogenesis of *Penicillium stoloniferum.* J Gen Microbiol 100: 59–64

Dieleman-van Zaayen A (1970) Means by which virus disease in cultivated mushrooms is spread, and methods to prevent and control it. Mushroom Growers' Assoc Bull 244: 158–178

Dieleman-van Zaayen A (1972) Intracellular appearance of mushroom virus in fruiting bodies and basidiospores of *Agaricus bisporus.* Virology 47: 94–104

Dieleman-van Zaayen A (1974) Stored spores of virus-infected mushrooms. Champignoncultuur 18: 221

Dieleman-van Zaayen A, Temmink JHM (1968) A virus disease of cultivated mushrooms in the Netherlands. Neth J Plant Pathol 74: 48–51

Ellis LF, Kleinschmidt WJ (1967) Virus-like particles of a fraction of statolon, a mould product. Nature (London) 215: 649–650

Fatemi F, Nelson RR (1977) Intra isolate heterokaryosis in *Pyricularia oryzae.* Phytopathology 67: 1523–1525

Ferault AC, Spire D, Rapilly F, Bertrandy J, Skajennikoff M, Bernaux P (1971) Observation de particules virales dans des souches de *Pyricularia oryzae.* Ann Phytopath 3: 267–269

Ferenczy L, Kevei F, Szegedi M (1976) Fusion of fungal protoplasts induced by polyethylene glycol. In: Peberdy JF, Rose AH, Rogers HJ, Cocking EC (eds) Microbial and plant protoplasts. Academic Press, New York, San Francisco, London, pp 177–187

Gandy DG (1960) A transmissible disease of cultivated mushrooms ("watery stipe"). Ann Appl Biol 48: 427–430

Gandy DG, Hollings M (1962) Die-back of mushrooms: a disease associated with a virus. Rep Glasshouse Crops Res Inst 1961, p 103–108

Giatcong P, Frederiksen RA (1969) Pathogenic variability and cytology of monoconidial subcultures of *Pyricularia oryzae.* Phytopathology 59: 1152–1157

Genovesi AG, Magill CW (1976) Heterokaryosis and parasexuality in *Pyricularia oryzae.* Can J Microbiol 22: 531–536

Hollings M (1962) Viruses associated with a die-back disease of cultivated mushroom. Nature (London) 196: 962–965

Hollings M (1967) Some aspects of virus disease in mushrooms. Mushroom Sci 6: 255–262

Hollings M (1978) Mycoviruses: Viruses that infect fungi. In: Lauffer MA, Maramorosch K, Bang FB, Smith KN (eds) Advances in virus research, vol 22. Academic Press, New York, San Francisco, London, pp.1–53

Hollings M, Gurney B (1973) In: Rep Glasshouse Crops Res Inst 1972, p 108

Hollings M, Stone OM (1971) Viruses that infect fungi. Annu Rev Phytopathol 9: 93–118

Hooper GR, Wood HA, Myers R, Bozarth RF (1972) Virus-like particles in *Penicillium brevicompactum* and *P. stoloniferum* hyphae and spores. Phytopathology 62: 823–825

Kneebone LR, Lockard JD, Hager RA (1962) Infectivity studies with X-disease. Mushroom Sci 5: 461–467

Koltin Y, Berick R, Stamberg J, Ben Shaul Y (1973) Virus-like particles and cytoplasmic inheritance of plaques in a higher fungus. Nature (London) New Biol 241:108–109

Lai HC, Zachariah K (1975) Detection of virus-like particles in coremia of *Penicillium claviforme.* Can J Genet Cytol 17; 525–533

Lapierre H, Faivre-Amiot A (1970) Presence de particules virales chez differentes souches de *Sclerotium cepivorum.* In: Proc 8th Cong Int Prot Plant, Paris, p 542

Lapierre H, Lemaire JM, Jouan B, Molin G (1970) Mise en evidence de particules virales associées à une perte de pathogénicité chez le Piétin-échaudage des céréales. C R Acad Sci Paris Ser D 271: 1833–1836

Lapierre H, Albouy J, Faivre-Amiot A, Molin G (1971) Mise en evidence de particules virales dans divers champignons du genre *Sclerotium.* CR Acad Sci Paris Ser D 272:2848–2851

Lapierre H, Faivre-Amiot A, Kusiak C, Molin G (1972) Particules de type viral associées au *Mycogone perniciosa* agent d'une des môles du champignon de couche. C R Acad Sci Paris Ser D 274: 1867–1870

Lapierre H, Faivre-Amiot A, Molin G (1973) Isolement de particules de type viral associées au *Verticillium fungicola:* agent d'une môle du champignon de couche. Ann Phytopathol 5: 323

Lecoq H, Delhotal P (1976) Etude des virus de *Colletotrichum lindemuthianum.* I Caracterisation des particules virales. Ann Phytopathol 8: 307–322

Lemke PA, Nash CH (1974) Fungal viruses. Bacteriol Rev 38: 29–56

Lemke PA, Nash CH, Pieper SW (1973) Lytic plaque formation and variation in virus titre among strains of *Penicillium chrysogenum.* J.Gen Microbiol 76: 265–275

Lhoas P (1970) Use of heterokaryosis to infect virus-free strains of *Aspergillus niger.* Aspergillus Newslett 11: 8–9

Lhoas P (1971a) Transmission of *ds*-RNA viruses to a strain of *Penicillium stoloniferum* through heterokaryosis. Nature (London) 230: 248–249

Lhoas P (1971b) Infection of protoplasts from *Penicillium stoloniferum* with *ds*-RNA viruses. J Gen Virol 13: 365–367

Lhoas P (1972) Mating pairs of *Saccharomyces cerevisiae* infected with *ds*-RNA viruses from *Aspergillus niger.* Nature (London) New Biol 236:86–87

MacDonald JG, Heath MC (1978) Rod shaped and spherical virus-like particles in cowpea rust fungus. Can J Bot 56: 963–975

Metitiri PO, Zachariah K (1972) Virus-like particles and inclusion bodies in penicillus cells of a mutant of *Penicillium.* J.Ultrastruct Res 40: 272–283

Milne RG, Luisoni E (1977) Rapid immune electron microscopy of virus preparations. In: Maramorosch K, Koprowski H (eds) Methods in virology, vol VI. Academic Press, New York, San Francisco, London, pp 265–281

Morgan DH, Oakley BA (1976) In: Rep John Innes Inst 1975, p 102

Pallett IH (1972) Production and regeneration of protoplasts from various fungi and their infection with fungal viruses. Abstr 3rd Symp Yeast Protoplasts, Salamanca, Spain, p 78

Pallett IH (1976) Interaction between fungi and their viruses. In: Peberdy JF, Rose AH, Rogers HJ, Cocking EC (eds) Microbial and plant protoplasts. Academic Press, New York, San Francisco, London, pp 107–124

Rawlinson CJ, Maclean DJ (1973) Virus-like particles in axenic cultures of *Puccinia graminis tritici.* Trans Br Mycol Soc 61: 590–593

Rawlinson CJ, Muthyalu G (1974) In: Rep Rothamsted Exp Stn, Harpenden, Engl, 1973, p 119

Rawlinson CJ, Hornby D, Pearson V, Carpenter JM (1973) Virus-like particles in the take-all fungus, *Gaeumannomyces graminis.* Ann Appl Biol 74: 197–209

Rawlinson CJ, Muthyalu G, Deacon JW (1977) Natural transmission of viruses in *Gaeumannomyces* and *Phialophora* spp. (Abstract) 2nd Int Mycol Congr , Abstracts, Tampa, Florida, p 558

Sansing GA, Detroy RW, Freer SN, Hesseltine CW (1973) Virus-like particles from conidia of *Penicillium* species. Appl Microbiol 26: 914–918

Schisler LC, Sinden JW, Sigel EM (1963) Transmission of a virus disease of mushrooms by infected spores. Phytopathology 53: 888

Schisler LC, Sinden JW, Sigel EM (1967) Etiology, symptomatology, and epidemiology of a virus disease of cultivated mushrooms. Phytopathology 57: 519–526

Shahriari H, Kirkham JB, Casselton LA (1973) Virus-like particles in the fungus *Coprinus lagopus.* Heredity 31: 428

Spire D, Ferault AC, Bertrandy J, Rapilly F, Skajennikoff M (1972) Particules de type viral dans un champignon hyperparasite: *Gonatobotrys.* Ann Phytopathol 4: 419

Takebe I (1975) The use of protoplasts in plant virology. Annu Rev Phytopathol 13: 105–125

Ushiyama R, Nakai Y (1975) Viruses assocated with Hymenomycetes. II. Presence of polyhedral virus-like particles in Shiitake mushroom, *Lentinus edodes.* Rep Tottori Mycol Inst Jpn 12: 53–60

Van Zaayen A (1976) Immunity of strains of *Agaricus bitorquis* to mushroom virus disease. Neth J Plant Pathol 82: 121–131

Wood HA, Bozarth RF (1973) Heterokaryon transfer of virus-like particles associated with a cytoplasmically inherited determinant in *Ustilago maydis.* Phytopathology 63: 1019–1021

Wood HA, Bozarth RF, Adler JP, MacKenzie DW (1974) Proteinaceous virus-like particles from
    an isolate of *Aspergillus flavus*. J Virol 13: 532–534
Yamashita S, Doi Y, Yora K (1973) Intracellular appearance of *Penicillium chrysogenum* virus.
    Virology 55: 445–452

Characterization of Fungal Viruses and Their Effect on the Host

Physicochemical Properties of Mycoviruses: An Overview

R.F. BOZARTH

Department of Life Sciences, Indiana State University, Terre Haute, IN 47809/USA

To comment on the physicochemical properties of all mycoviruses and viruslike particles (VLPs) isolated from fungi in twenty minutes is a formidable task, for many shapes, sizes, and chemical compositions have been documented or postulated. The task is greatly simplified by the fact that only a few mycoviruses have been well characterized and most of these have common properties. In a recent review of physicochemical properties of mycoviruses, only those from 17 fungus species were included (Bozarth, 1979) since most of those described from other species (Lemke, 1979) are known only from electron microscopic investigations.

In this brief overview, I shall attempt to make some generalizations based on the properties of well-characterized mycoviruses. All mycoviruses do not have all of the general properties described in the following pages, but in arriving at a general impression of mycoviruses, these properties seem to prevail, at least for those which have been studied beyond the mere observation by electron microscopy.

1 Isometric Particles 25–50 nm Diameter

All but a few of the VLPs isolated from fungi are isometric and have diameters within the range of 25–48 nm (Fig. 1). If one accepts reported diameters at face value, there is a continuum of particle diameters over this range. Table 1 lists the diameters of the most thoroughly studied mycoviruses arranged in order of reported size..

The size range, 35–40 nm, is most typical for fungal viruses. Over one-half of the reported diameters fall within this range. The fact that some viruses are reported as having different diameters by different laboratories emphasises the difficulty in ascertaining the exact particle diameter. For instance, *Penicillium chrysogenum* virus, the most thoroughly studied fungal virus, has been reported to be 35 nm diameter (Buck and Girvan, 1977) and 40 nm (Wood and Bozarth, 1972).

The next most frequently reported diameter is 27–34 nm. Again the difficulty in determining the exact size is illustrated by the fact that in negatively stained preparations of *P. stoloniferum* virus complex, particles penetrated by the stain measure 31 nm, whereas particles not penetrated by the stain measure 34 nm diameter (Bozarth et al., 1971). In *Aspergillus flavus* VLPs measured 27 nm if measured singly or 30 nm if measured from packed arrays in the same electron micrographs (Table 1, Wood et al., 1974).

Mycoviruses, when negatively stained with either uranyl acetate or sodium phosphotungstate, tend to be penetrated by the stain. This is especially true of VLPs in the size range 35–40 nm, but the larger particles from *Helminthosporium maydis* (48 nm) are

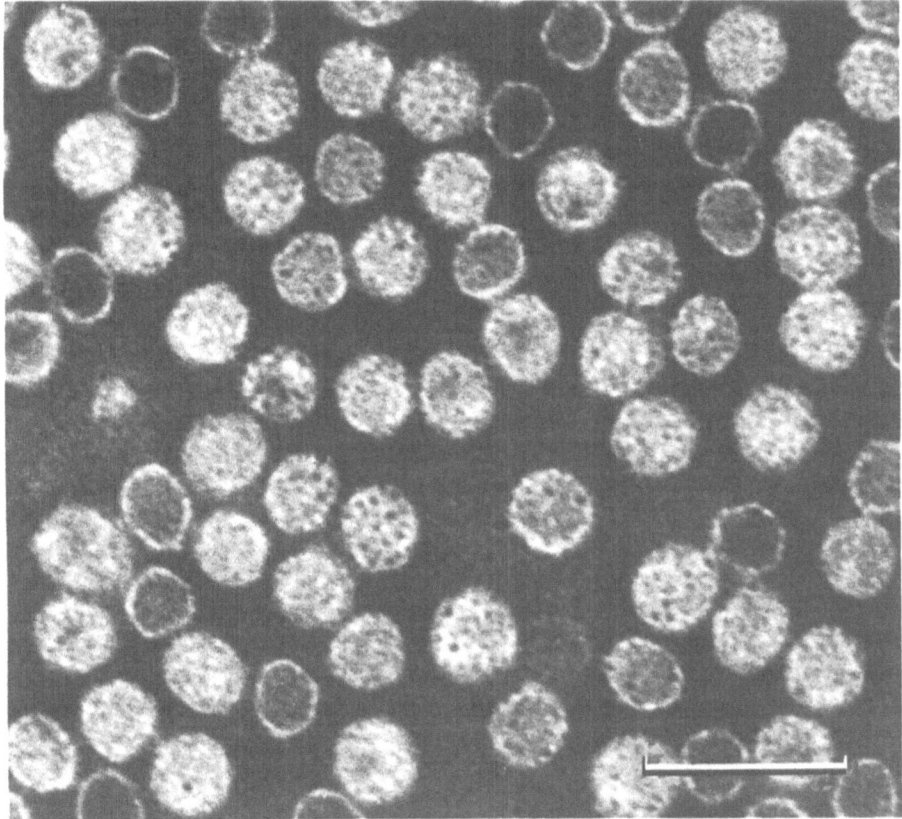

Fig. 1. Electron micrograph of negatively-stained mycovirus from *Thielaviopsis basicola.* Bar equals 100 nm

also usually penetrated (Bozarth, 1977). One cannot rely on stain penetration to distinguish between particles which contain nucleic acid (full particles) and those which do not contain nucleic acid (empty particles). While susceptibility to stain penetration may be regarded as a general property of mycoviruses, it also contributes to the difficulty in determining particle diameters, for as noted above, penetrated and unpenetrated particles frequently have different measured diameters.

The exact particle diameters are difficult and perhaps impossible to determine. Electron micrographs of negatively stained VLPs are most often used to determine particle diameter, but images produced with different staining materials are quite variable. Problems encountered in negative staining of virus particles have been recently discussed by Milne and Luisoni (1977) and their conclusions are in agreement with my own observations. Particles negatively stained with 2% sodium phosphotungstate adjusted to pH 7.0 often appear flattened and distorted. One can usually produce some micrographs in which the particles do not appear flattened and those are used to measure particle diameter, but this is a very subjective procedure. One percent uranyl acetate usually produces better images provided the buffer salts are removed by washing the grid to which the sample has been applied with H_2O before touching the grid to the solution of uranyl acetate.

Table 1. Some reported particle diameters of mycoviruses and VLPs from fungi

Fungus species	Diameter	Reference
Penicillium funiculosum	25–30	Banks et al. (1968)
Agaricus bisporus	25	Hollings (1962)
" "	25	Molin and Lapierre (1973)
" "	26	Marino et al. (1976)
Gaeumannomyces graminis	27	Rawlinson et al. (1973)
Aspergillus flavus		
(measured singly)	27	Wood et al. (1974)
(measured from packed arrays)	30	"
Agaricus bisporus	29	Hollings (1962)
Colletotrichum lindemuthianum	30	Rawlinson et al. (1975)
Penicillium stoloniferum		
(particles penetrated by stain)	31	Bozarth et al. (1971)
(particles not penetrated by stain)	34	"
Agaricus bisporus	34	Molin and Lapierre (1973)
Gaeumannomyces graminis	35	Rawlinson et al. (1973)
Penicillium chrysogenum	35	Buck and Girvan (1977)
Penicillium cyaneo-fulvum	35	Buck and Girvan (1977)
Agaricus bisporus	35	Marino et al. (1976)
Saccharomyces cerevisiae	39	Herring and Bevan (1974)
" "	40	Adler et al. (1976)
Penicillium chrysogenum	40	Wood and Bozarth (1972)
Gaeumannomyces graminis	40	Almond et al. (1977)
Penicillium brevi-compactum	40	Wood et al. (1971)
Allomyces arbuscula	40	Khandjian et al. (1977)
Thielaviopsis basicola	40	Bozarth and Goenaga (1977)
Aspergillus foetidus	40–42	Banks et al. (1970)
Ustilago maydis	41	Wood and Bozarth (1973)
Helminthosporium maydis	48	Bozarth (1977)

Even with well-formed images, difficulties are encountered in determining the exact location of the edge of the particle. Overfocusing to produce a clear edge does not solve the problem. The common practice of determining the average particle diameter by measuring packed arrays may be deceptive because the degree of overlapping of particles cannot be ascertained.

The buffer used to suspend virions or VLPs almost certainly affects the appearance of the particle in electron micrographs, but this parameter has not been explored with respect to fungal viruses.

In spite of these difficulties and the great variability of reported particle diameters, most isometric viruses from fungi appear to be grouped in the sizes noted. The diameter most often found (35–40 nm) is uncommon for isometric viruses in general.

2 Capsid Structure of T = 1

On the basis of chemical analysis carried out in the laboratory of K.W. Buck, five mycoviruses have been postulated to have a capsid structure conforming to the geometric de-
sign T = 1. These are the S and F viruses of *A. foetidus* (Buck and Ratti, 1975), *P.
chrysogenum* virus (Buck and Girvan, 1977), and the S and F viruses of *P. stoloniferum*
(Buck and Kempson-Jones, 1974).

No ultrastructural analyses of fungal viruses have been published, but it may be of
interest to note that the appearance of most fungal viruses in negatively stained preparations is distinctly different from the appearance of the slightly smaller isometric plant
viruses, all of which when studied have been found to have structural symmetry of T = 3.

3 Genomes of Segmented dsRNA

Mycoviruses typically have genomes of dsRNA which are segmented into discrete pieces.
These pieces have been referred to as segments (Bozarth and Harley, 1976), classes
(Wood and Bozarth, 1972), components (Wood and Bozarth, 1973; Buck and Girvan,
1977), or species (Burnett et al., 1975).

The development of polyacrylamide gel electrophoresis (PAGE) for the analysis of
RNA (Loening and Ingle, 1967) just preceded the surge of interest in fungal viruses
which followed the discovery that *P. stoloniferum* contained a dsRNA virus (Ellis and
Kleinschmidt, 1967). The application of PAGE to the analysis of dsRNA segments was
initiated by Shatkin et al. (1968) who showed that the mobility of three groups of
dsRNA segments in reovirus was approximately inversely proportional to their molecular weights. These segments became the standards for molecular weight determinations
of mycovirus dsRNA segments and provided accurate results within the limits of 0.8
to 2.5×10^6 daltons. However, attempts to extrapolate beyond these limits led to frus-
tration. The extrapolated values did not correspond to results obtained with an equally
valid set of standards determined by analysis of the dsRNA of the $\phi6$ bacteriophage
(Semancik et al., 1973). This led to a reevaluation of the relationships between gel mobility and molecular weight of the dsRNA (Bozarth and Harley, 1976). When the dsRNA
segments of *P. chrysogenum* virus, *P. stoloniferum* virus, *H. maydis* virus, $\phi6$ bacteriophage, and reovirus were simultaneously electrophoresed on the same gels, the relationship of mobility to log molecular weight as determined by contour length in electron
micrographs was found to be a smooth curve with all points falling very close to the
line. The range of molecular weights which could be analyzed by this method was extended to cover 0.45 to 6.3×10^6 daltons (Bozarth and Harley, 1976). This relationship has since been confirmed for dsRNA (Buck and Ratti, 1977) and for ssRNA (Kaper and Diaz-Ruiz, 1977). Some early values based on extrapolations have been cor-
rected (Bozarth, 1978). Other determinations based on extrapolation are almost certainly erroneous (Dunkle, 1974; Khandjian et al., 1977; Sanderlin and Ghabrial, 1978).

Most reports of molecular weights of dsRNA segments are reported to three significant figures, a questionable practice since the analysis of the standards upon which
they are based is at best two significant figures. In analysis by relative migration (PAGE),

the species being analyzed can have no greater accuracy than the standards upon which
the analysis is based. However, this practice can be defended since it affords an estimate
of *relative* sizes where small differences in molecular weights exist — differences so
small that they are resolved only by PAGE.

The tendency of mycoviruses to have genomes of dsRNA and the relatively easy
analysis of dsRNA by PAGE have provided the basis for analyses and study of genetic
and biological characteristics of several mycoviruses in which the virus purification was
quite difficult. The viruses of *Saccharomyces cerevisiae* and *Ustilago maydis* are both
difficult to produce in a highly purified state (Wood and Bozarth, 1973; Adler et al.,
1976), yet direct extraction and analyses of dsRNA have been the basis for excellent
genetical studies with both viruses (Koltin and Day, 1976a; Koltin, 1978; see chapter
9 by Koltin in this book). A virus of *Endothia parasitica* has been recognized, studied
biologically, and partially purified as a result of the fact that its genome contains dsRNA
(Day et al., 1977).

All of the isometric mycoviruses studied to date contain segmented dsRNA genomes. These include the viruses of *Agaricus bisporus* (Marino et al., 1976), *Allomyces
arbuscula* (Khandjian et al., 1977), *Aspergillus foetidus* (Ratti and Buck, 1972; Buck
and Ratti, 1975, 1977), *A. niger* (Buck et al., 1973a; Buck and Ratti, 1977), *Colletotrichum lindemuthianum* (Rawlinson et al., 1975), *Gaeumannomyces graminis* (Rawlinson et al., 1973), *Helminthosporium maydis* (Bozarth, 1977), *H. victoriae* (Sanderlin and Ghabrial, 1978), *Penicillium chrysogenum* (Wood and Bozarth, 1972; Buck and
Girvan, 1977), *P. brevi-compactum* (Wood et al., 1971), *P. funiculosum* (Lampson et
al., 1967), *P. stoloniferum* (Banks et al., 1968; Bozarth and Wood, 1970; Buck and
Kempson-Jones, 1970, 1974; Bozarth et al., 1971), *Periconia circinata* (Dunkle, 1974),
Saccharomyces cerevisiae (Berry and Bevan, 1972; Bevan et al., 1973; Buck et al., 1973;
Vodkin and Fink, 1973; Herring and Bevan, 1974; Adler, 1975; Adler et al., 1976;
Wickner, 1976), *Thielaviopsis basicola* (Bozarth and Goenaga, 1977), and *Ustilago may-
dis* (Wood and Bozarth, 1973; Koltin and Day, 1976a; Lentz, 1977; Koltin, 1978;
Bozarth and Lentz, 1978).

Some isometric mycoviruses have also been reported to contain single-stranded RNA.
Buck and Kempson-Jones (1973) reported that all combinations of single- and double-
stranded RNA are contained in some particles of *P. stoloniferum* virus which are separable by sucrose-density-gradient centrifugation. The mycovirus of *H. maydis* has a
major sedimenting component which contains dsRNA and a slower sedimenting "middle" component which has not been thoroughly analyzed. Its RNA may prove to be
single-stranded (Bozarth, 1977).

4 Isometric Mycoviruses Typically Contain a Single Piece of dsRNA per Virion

This fact was first appreciated when results of chemical and physical analysis suggested
that the *P. chrysogenum* virus contained approximately 15% RNA rather than the 45%
calculated if each virus contained all of the three segments of dsRNA found in purified
preparations of the virus (Wood and Bozarth, 1972). Partial degradation and electron
microscopy (Fig. 2) revealed that none of over 1000 particles examined had more than

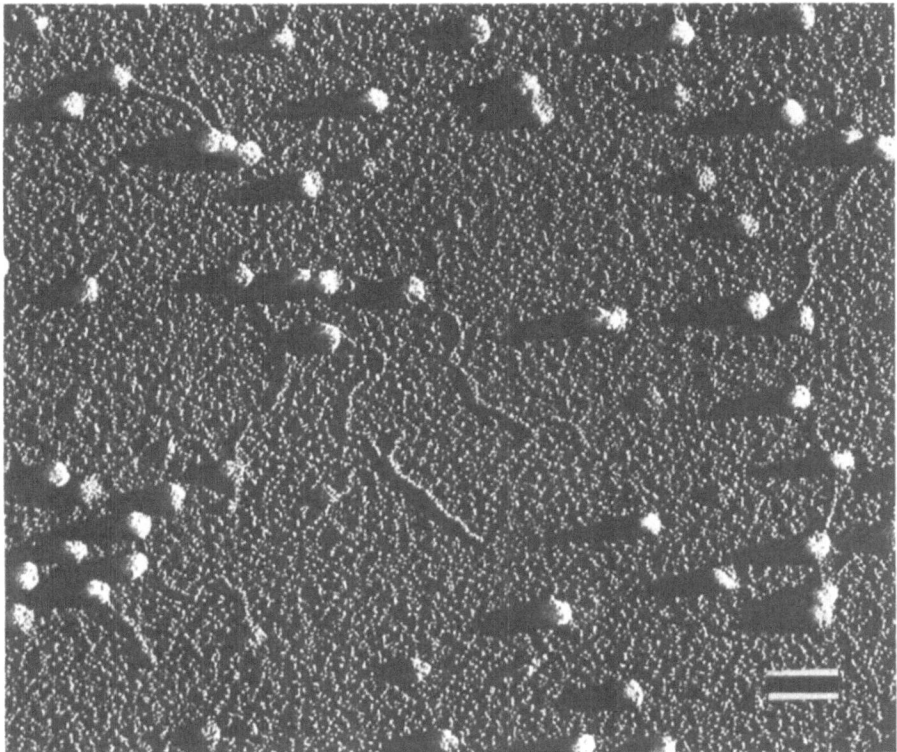

Fig. 2. Electron micrograph of partially degraded virions from *Penicillium chrysogenum* shadowed with PdPt. A single piece of dsRNA is observed extending from otherwise intact nucleocapsids. Bar equals 100 nm

a single dsRNA segment extending from the partially degraded capsid (Wood and Bozarth, 1972). Furthermore, sucrose-density-gradient centrifugation and analysis of the leading, middle, and trailing portions of the sedimenting bands confirmed that the fastest sedimenting portion contained the largest piece of dsRNA and the middle and slower sedimenting portions contained the respective segments of dsRNA (Buck and Girvan, 1977).

Other viruses which have a single piece of dsRNA per virion include *H. maydis,* component C (Bozarth and Harley, 1976; Bozarth, 1977), and *H. victoriae* Sanderlin and Ghabrial, 1978). The viruses of *P. stoloniferum* and *A. foetidus* both appear to have some particles with a single piece of dsRNA and others with combinations of pieces (Buck and Ratti, 1975; Buck and Kempson-Jones, 1973).

5 Multiple Infections Are Common

Among the mycoviruses characterized thus far, those from *P. stoloniferum* (Bozarth et al., 1971; Buck and Kempson-Jones, 1970, 1973, 1974), *A. foetidus* (Buck and Ratti, 1977), *P. circinata* (Dunkle, 1974), *T. basicola* (Bozarth and Goenaga, 1977), *H. victo-*

riae (Sanderlin and Ghabrial, 1978), and *G. graminis* (Rawlinson et al., 1973; Almond et al., 1977) occur as mixed infections of two to several serologically distinct viruses. *T. basicola* provides a good example. At least eight sedimenting components are resolvable by repeated fractionation in sucrose-density gradients (Fig. 3). This complex

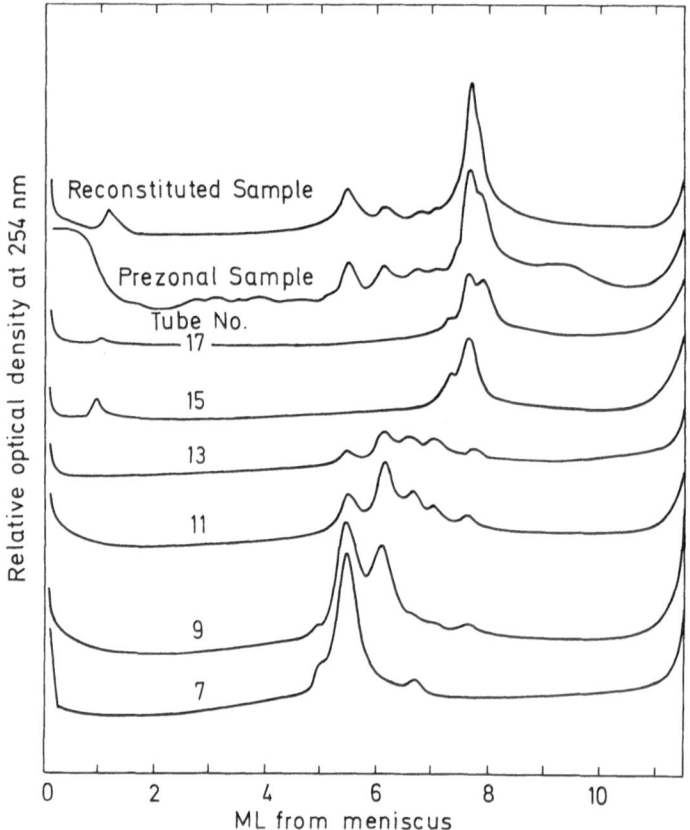

Fig. 3. Sucrose-density-gradient analysis of the components of a complex of VLPs from *Thielaviopsis basicola*. The virus complex (prezonal sample) was fractionated by zonal centrifugation into zones with the indicated tube number. Each tube was then analyzed on analytical sucrose-density gradients as indicated for tubes 7–17. When equal aliquots of each fraction were combined, the reconstituted sample appeared quite similar to the original prezonal sample. From these experiments, it was deduced that there were at least eight sedimenting components. (From Bozarth and Goenaga, 1977, with permission of the publisher)

contains five serologically distinct capsids (Fig. 4) which may be correlated with five segments of dsRNA separable by PAGE. Almond et al. (1977) purified virus from a number of isolates of *G. graminis*. These isolates contain serologically distinct nucleocapsids. One of these isolates, 45/10, contained five serological components and twelve segments of dsRNA.

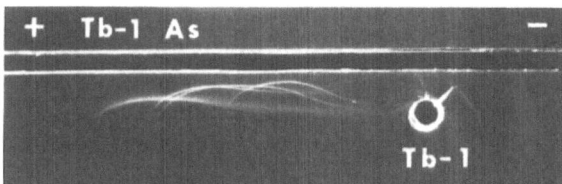

Fig. 4. Immunoelectrophoresis of the VLP complex from *Thielaviopsis basicola* (Tb-1). Electrophoresis of the complex (see Fig. 3) was for 1.5 h at 5mA in 0.1 M potassium phosphate buffer and the serological reactions were developed 36 h before the picture was taken. Five distinct capsid antigens are indicated by the reaction. (From Bozarth and Goenaga, 1977, with permission of the publisher)

6 Genomic Masking or Mixing of Capsid Proteins is Not Known to Occur

Although multiple infections are common in fungi, genomic masking and mixing of capsid proteins are not known to occur. The best studied example is the complex of PsV-S and PsV-F viruses from *P. stoloniferum*. Although it has been shown that both viruses multiply in the same cells (Bozarth et al., 1971; Adler and Mackenzie, 1972), immunoelectrophoresis of capsids revealed no capsid proteins migrating at an intermediate rate, and double-diffusion serological tests showed no spur formation between reaction lines of PsV-S and PsV-F. Furthermore, the capsids of PsV-S contain two dsRNA segments (individually encapsidated) and those of PsV-F contain three segments (Fig. 5). The failure of the dsRNA segments to become encapsidated in the

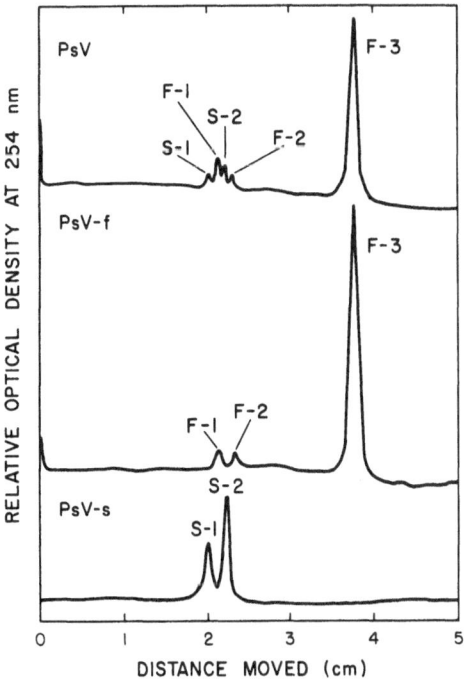

Fig. 5. Polyacrylamide gel electrophoresis scanning patterns of dsRNA segments from the complex of viruses in *Penicillium stoloniferum* (ATCC 14586). PsV-F contains three dsRNA segments whereas PsV-S contains only two. The segments labeled S1, S2, F1, and F2 are very close to the same molecular weight and could very easily be encapsidated in the "wrong" capsid if size of the RNA were the only factor. (From Bozarth et al., 1971, with permission of the publisher)

"wrong" capsid seems remarkable in view of their similarity in molecular weight, and suggests a close correlation of capsid protein synthesis with RNA replication within the doubly-infected cells.

A somewhat similar situation occurs in *A. niger* which also contains electrophoretically fast and slow species of mycoviruses. A segment of 2.31×10^6 daltons was common to both the fast and slow species. This was thought to represent the first case of genomic masking in fungal viruses (Buck et al., 1973a). Later analysis (K.W. Buck, personal communication) indicated that the RNAs had different nucleotide base ratios and were therefore distinct RNA molecules.

7 Serologically Related Mycoviruses Are Found in Different Fungal Species

In spite of the relatively small number of mycoviruses which have been studied in detail, there are four groups of related species – one with three and three with two members.

The first group consists of the mycoviruses from *P. chrysogenum* (Wood and Bozarth, 1972), *P. brevi-compactum* (Wood et al., 1971), and *P. cyaneo-fulvum* (Buck and Girvan, 1977). The mycoviruses of *P. brevicompactum* and *P. chrysogenum* produce reciprocal spurs in heterologous serological tests. Their dsRNA segments have the same molecular weight, but there is a difference in the relative quantity of the three segments in purified preparations. *P. cyaneo-fulvum* has been shown to be serologically related to *P. chrysogenum*, and in addition to the "normal" three dsRNA segments, it has a fourth segment of smaller molecular weight (Buck and Girvan, 1977).

Serologically related mycoviruses have been isolated from *A. foetidus* (Ratti and Buck, 1972) and *A. niger* (Buck et al., 1973a). Both the slow and the fast viruses from these species are related to the respective virus from the other species.

The *P. stoloniferum*-S virus is serologically related to a virus extracted from *Diplocarpon rosae* (Bozarth et al., 1972).

8 Other Forms of Mycovirus

Not all viruslike particles observed from fungi fit into the above generalizations. Of these, the best described is a bacilliform type which was among the first viruses isolated from *A. bisporus* by Hollings (1962). They are virtually identical in appearance to the alfalfa mosaic virus, and it has been proposed (Hollings, 1978) that they be grouped taxonomically with alfalfa mosaic virus. Molin and Lapierre (1973) obtained similar particles from mushrooms grown in France and reported that they contained ssRNA. Moyer and Smith (1976) produced an antiserum to similar particles isolated from *A. bisporus* in Pennsylvania but listed no other properties, and apparently their virus preparation was not tested against the antiserum of Hollings.

An interesting recent development is the isolation from *A. bisporus* of pleomorphic particles which appear to have a membrane by Lesemann and Koenig (1977). These particles (Fig. 6) were associated with, but not proven to be the cause of a disease of

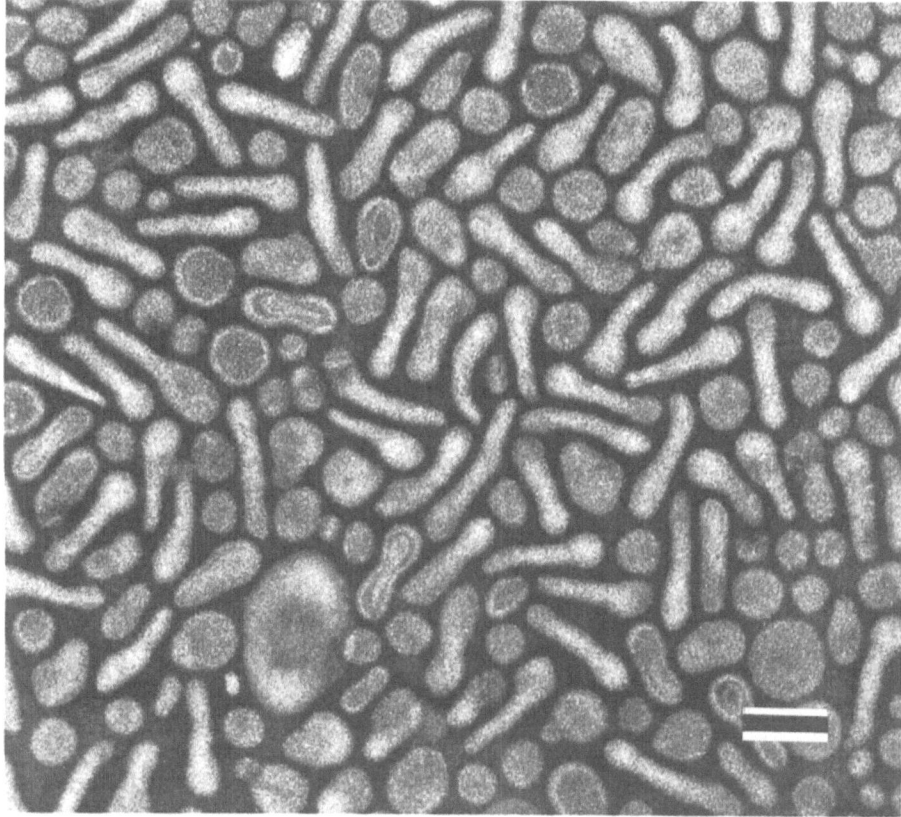

Fig. 6. Electron micrograph of partially purified preparations of clubshaped particles isolated from *Agaricus bisporus* by Lesemann and Koenig (1977). Bar equals 100 nm

mushrooms. They are similar in appearance to particles observed earlier in thin sections of diseased mushrooms by Albouy et al. (1973). Day et al. (1977) reported the isolation of dsRNA from hypovirulent strains of *Endothia parasitica*. Further analysis (J.A. Dodds, personal communication) indicates that this dsRNA is associated with particles which are similar in appearance to those isolated by Lesemann and Koenig. Particles of this type would not have been detected by the procedures of most investigators who concentrated on detecting particles of uniform appearance in the electron miscroscope. Such particles may indeed be widespread in fungi.

Rod-shaped VLPs may qualify as mycoviruses, but definitive information based on studies of purified preparations has not been presented. Dieleman-van Zaayen (1967) and Dieleman-van Zaayen et al. (1970) reported particles 17 × 350 nm from *Plicaria fulva*, Lapierre et al. (1972) reported particles 18 × 120 nm from *Mycogone perniciosa*, Inoue (1970) reported particles 28 × 300 nm from *Lentinus cdodes*, and Blattný et al. (1973) reported particles 22–28 × 119 nm from *Armillariella mellea*.

Excellent electron micrographs of VLPs which resemble herpes virus have been produced from thin sections of *Thraustochytrium* sp. by Kazama and Schornstein (1972,

1973) and Pollak (1975), but no purified preparations have been made and the similarity with herpes virus is based on interpretations from thin sections. Likewise the 200 nm particles from *Aphelidium* sp. reported by Schnepf et al. (1970, 1971), the 60 nm particles from *Neurospora* reported by Tuveson and Peterson (1972), the 120–170 nm particles from *Neurospora* reported by Küntzel et al. (1973), the 130 nm particles from *Schizophyllum commune* reported by Koltin et al. (1973), and the 18–20 nm particles from *Neurospora* reported by Tuveson and Peterson (1972) and Tuveson et al. (1975) have not been adequately investigated.

Finally, acceptance of reports of VLPs from fungi which resemble known bacteriophage (Volkoff and Walters, 1970; Tikhonenko, 1978) must be withheld until it is conclusively proven that these particles did not arise from contaminating bacteria.

Summary

The physicochemical properties of mycoviruses from 14 fungal species have been studied in detail. They are similar in properties and may be considered a homogeneous group. They fall into three groups of 30–34, 36–40, and 48 nm isometric particles which, where studied, have a single major capsid peptide arranged in T = 1 symmetry containing 60 or 120 subunits per particle. All have genomes of dsRNA which are usually segmented. Segments are usually singly encapsidated into otherwise identical capsids, resulting in a family of particles with slightly different sedimentation velocities or densities in CsCl. Similar particles have been reported on the basis of electron microscopic evidence from many other species.

Serologically related viruses have been identified in *Aspergillus foetidus* and *A. niger; Penicillium stoloniferum* and *Diplocarpon rosae;* and in *P. chrysogenum, P. cyaneo-fulvum,* and *P. brevicompactum.*

Multiple infections with serologically distinct species are common. In *P. stoloniferum,* the serologically distinct viruses PsV-S and PsV-F were found to multiply in the same cells and always at the same relative rate. There was no evidence of mixing of genomic segments or capsid peptides. The two viruses of *Aspergillus foetidus,* Af-S and Af-F, appear to be similar to the *P. stoloniferum* viruses in this respect. Strains of *Thielaviopsis basicola* and *Gaeumannomyces graminis* have been reported to contain five distinct nucleoprotein species.

Other shapes of particles have been reported, but their descriptions are less complete. These include the short rods from *Agaricus bisporus* and filamentous rods from *Peziza ostracoderma;* large herpes-type particles from *Thraustochytrium* species; large 300 nm lipid-containing particles and small 18–20 nm particles which aggregate in pairs from *Neurospora crassa;* and a T-2 type bacteriophage from *Penicillium brevicompactum.*

Acknowledgments. The author expresses his thanks and appreciation for grants from the Deutscher Akademischer Austauschdienst and the Indiana State University which made the presentation of this paper possible. Research by the author was carried out with assistance from the National Institute of Health (12–14–100134) and the U.S. Department of Agriculture. The electron micrograph shown in Figure 6 was by D.E. Lesemann.

References

Adler JP (1975) Viruses of yeast. Dev Ind Microbiol 16: 152–157

Adler JP, Mackenzie DW (1972) Intrahyphal localizations of *Penicillium stoloniferum* viruses by fluorescent antibody. Abstr Annu Meet Am Soc Microbiol, p 68

Adler J, Wood HA, Bozarth RF (1976) Virus-like particles from killer, neutral, and sensitive strains of *Saccharomyces cerevisiae.* J. Virol 17: 472–476

Albouy MM, Lapierre H, Molin G (1973) Mise en évidence d'un nouveau type de particules dans des hyphes de carpophores d'*Agaricus bisporus.* C R Acad Sci (Paris) Ser D 276:2805–2807

Almond MR, Buck KW, Rawlinson CJ (1977) Viruses of *Gaeumannomyces graminis* var. *tritici.* Abstr 2nd Int Mycol Congr (Tampa). p 104

Banks GT, Buck KW, Chain EB, Himmelweit F, Marks JE, Tyler JM, Hollings M, Last FT, Stone OM (1968) Viruses in fungi and interferon stimulation. Nature (London) 218: 542–545

Banks GT, Buck KW, Chain EB, Darbyshire JE, Himmelweit F, Ratti G, Sharpe TJ, Planterose DN (1970) Antiviral activity of double-stranded RNA from a virus isolated from *Aspergillus foetidus.* Nature (London) 227: 505–507

Berry EA, Bevan EA (1972) A new species of double stranded RNA from yeast. Nature (London) 239: 279–280

Bevan EA, Herring AJ, Mitchell DJ (1973) Preliminary characterization of two species of *ds*RNA in yeast and their relationship to killer character. Nature New Biol (London) 458: 81–86

Blattný C, Králik O, Veselsky J, Kasala B, Herzova H (1973) Particles resembling virions accompanying the proliferation of agaric mushrooms. Ceská Mykol 27: 1–5

Bozarth RF (1977) Biophysical and biochemical characterization of viruslike particles containing a high molecular weight ds-RNA from *Helminthosporium maydis.* Virology 80: 149–157

Bozarth RF (1979) The physico-chemical properties of dsRNA mycoviruses. In: Viruses and plasmids in fungi. Lemke PA (ed). Marcel Dekker Inc, New York, in press

Bozarth RF, Goenaga A (1977) A complex of virus-like particles from *Thielaviopsis basicola.* J Virol 24: 846–849

Bozarth RF, Harley EH (1976) The electrophoretic mobility of double-stranded RNA in polyacrylamide gels as a function of molecular weight. Biochim Biophys Acta 432: 329–335

Bozarth RF, Lentz ET (1978) Physico-chemical properties and serology of *Ustilago maydis* viruses P1 and P4. Abstr 4th Int Congr Virol (The Hague, The Netherlands), p 172

Bozarth RF, Wood HA (1970) Purification and properties of two viruses from *Penicillium stoloniferum.* Phytopathology (Abstr) 60: 1285

Bozarth RF, Wood HA, Mandelbrot A (1971) The *Penicillium stoloniferum* virus complex: two similar double-stranded RNA virus-like particles in a single cell. Virology 45: 516–523

Bozarth RF, Wood HA, Goenaga A (1972) Virus-like particles from a culture of *Diplocarpon rosae.* Phytopathology (Abstr) 62: 493

Buck KW, Girvan RF (1977) Comparison of the biophysical and biochemical properties of *Penicillium cyaneo-fulvum* virus and *Penicillium chrysogenum* virus. J Gen Virol 34: 145–154

Buck KW, Kempson-Jones GF (1970) Three types of virus particles in *Penicillium stoloniferum.* Nature (London) 225: 945–946

Buck KW, Kempson-Jones, GF (1973) Biophysical properties of *Penicillium stoloniferum* virus S. J Gen Virol 18: 223–235

Buck KW, Kempson-Jones GF (1974) Capsid polypeptides of two viruses isolated from *Penicillium stoloniferum.* J Gen Virol 22: 441–445

Buck KW, Ratti G (1975) Biophysical and biochemical properties of two viruses isolated from *Aspergillus foetidus.* J Gen Virol 27: 211–224

Buck KW, Ratti G (1977) Molecular weight of double-stranded RNA: a re-examination of *Aspergillus foetidus* virus S RNA components. J Gen Virol 37: 215–219

Buck KW, Girvan RF, Ratti G (1973a) Two serologically distinct double-stranded ribonucleic acid viruses isolated from *Aspergillus niger.* Biochem Soc Trans 1: 1138–1140

Buck KW, Lhoas P, Border DJ, Street BK (1973b) Virus particles in yeast. Biochem Soc Trans 1: 1141–1142

Burnett JP, Frank BH, Douthart RJ (1975) Ionic-strength effects on stability and conformation of *Penicillium chrysogenum* mycophage double-stranded RNA. Nucleic Acid Res 2: 759–771

Day PR, Dodds JA, Elliston JE, Jaynes RA, Anagnostakis SL (1977) Double-stranded RNA in *Endothia parasitica.* Phytopathology 67: 1393–1396

Dieleman-van Zaayen A (1967) Virus-like particles in a weed mould growing on mushroom trays. Nature (London) 216: 595–596

Dieleman-van Zaayen A, Igesz O, Finch JT (1970) Intracellular appearance and some morphological features of virus-like particles in an ascomycete fungus. Virology 42: 534–537

Dunkle LD (1974) Double-stranded RNA mycovirus in *Periconia circinata*. Physiol Plant Pathol 4: 107–116

Ellis LF, Kleinschmidt WJ (1967) Virus-like particles of a fraction of statolon, a mould product. Nature (London) 215: 649–650

Herring AJ, Bevan EA (1974) Virus-like particles associated with the double-stranded-RNA species found in killer and sensitive strains of the yeast *Saccharomyces cerevisiae*. J Gen Virol 22: 387–394

Hollings M (1962) Viruses associated with a die-back disease of cultivated mushroom. Nature (London) 196: 962–965

Hollings M (1978) Mycoviruses: viruses that infect fungi. In: Lauffer MA, Bang FB, Maramorosch K, Smith KM (eds) Advances in virus research, vol 22. Academic Press, New York, pp 1–53

Inoue T (1970) Virus-like particles in *L. edodes*. Mushrooms 2: 18–22

Kaper JM, Diaz-Ruiz JR (1977) Molecular weights of the double-stranded RNAs of cucumber mosaic virus strain S and its associated RNA 5. Virology 80: 214–217

Kazama FY, Schornstein KL (1972) Herpes-type virus particles associated with a fungus. Science 177: 696–697

Kazama FY, Schornstein KL (1973) Ultrastructure of a fungus herpes-type virus. Virology 52: 478–487

Khandjian EW, Turian G, Eisen H (1977) Characterization of the RNA mycovirus infecting *Allomyces arbuscula*. J Gen Virol 35: 415–424

Koltin Y (1978) Virus-like particles in *Ustilago maydis:* mutants with partial genomes. Genetics 86: 527–534

Koltin Y, Day PR (1976a) Inheritance of killer phenotypes and double-stranded RNA in *Ustilago maydis*. Proc Natl Acad Sci USA 73: 594–598

Koltin Y, Day PR (1976b) Suppression of the killer phenotype in *Ustilago maydis*. Genetics 82: 629–637

Koltin Y, Berick R, Stamberg J, Ben-Shaul Y (1973) Virus-like particles and cytoplasmic inheritance of plaques in a higher fungus. Nature New Biol 241: 108–109

Küntzel H, Barath Z, Ali I, Kind J, Althaus NH (1973) Virus-like particles in an extranuclear mutant of *Neurospora crassa*. Proc Natl Acad Sci USA 70: 1574–1578

Lampson GP, Tytell AA, Field AK, Nemes MM, Hilleman MR (1967) Inducers of interferon and host resistance. 1. Double-stranded RNA from extracts of *Penicillium funiculosum*. Proc Natl Acad Sci USA 58: 782–789

Lapierre H, Faivre-Amiot A, Kusiak C, Molin G (1972) Particules de type viral associées au *Mycogone perniciosa* Magnus, agent d'une des môles du champignon de couche. C R Acad Sci (Paris) Ser D 274: 1867–1870

Lemke PA (ed) (1979) Viruses and plasmids in fungi. Marcel Dekker Inc, New York, in press

Lentz E (1977) Physical, molecular, and serological characteristics of the virus-like particles of the corn smut fungus: *Ustilago maydis*. M A Thesis, Indiana State University

Lesemann DE, Koenig R (1977) Association of clubshaped virus-like particles with a severe disease of *Agaricus bisporus*. Phytopathol Z 89: 161–169

Loening UE, Ingle J (1967) Diversity of RNA components in green plant tissues. Nature (London) 215: 363–367

Marino R, Saksena KN, Schuler M, Mayfield JE, Lemke PA (1976) Double-stranded RNA from *Agaricus bisporus*. Appl Environ Microbiol 31: 433–438

Milne RG, Luisoni E (1977) Rapid immune electron microscopy of virus preparations. In: Maramorosch K, Koprowski H (eds) Methods in virology, vol 6. Academic Press, New York, and London, pp 265–281

Molin G, Lapierre H (1973) L'acide nucléique des virus de champignons: cas des virus de l'*Agaricus bisporus*. Ann Phytopathol 5: 233–240

Moyer JW, Smith SH (1976) Partial purification and antiserum production to the 19 X 50-nm mushroom virus particle. Phytopathology 66: 1260–1261

Pollak A (1975) The induction of virus-like particles in the marine phycomycete, *Thraustochytrium aureum*. Abstr Annu Meet Am Soc Microbiol, p 190

Ratti G, Buck KW (1972) Virus particles in *Aspergillus foetidus:* a multicomponent system. J Gen Virol 14: 165–175

Rawlinson CJ, Hornby D, Pearson V, Carpenter JM (1973) Virus-like particles in the take-all fungus, *Gaeumannomyces graminis.* Ann Appl Biol 74: 197–209

Rawlinson CJ, Carpenter JM, Muthyalu G (1975) Double-stranded RNA virus in *Colletotrichum lindemuthianum.* Trans.Br Mycol Soc 65: 305–341

Sandèrlin RS, Ghabrial SA (1978) Physicochemical properties of two distinct types of virus-like particles from *Helminthosporium victoriae.* Virology 87: 142–151

Schnepf E, Soeder CJ, Hegewald E (1970) Polyhedral virus-like particles lysing the aquatic phycomycete *Aphelidium* sp., a parasite of the green alga *Scenedesmus alatus.* Virology 42: 482–487

Schnepf E, Hegewald E, Soeder CJ (1971) Elektronenmikroskopische Beobachtungen an Parasiten aus *Scenedesmus*-Massenkulturen. 2. Über Entwicklung und Parasit-Wirt Kontakt von *Aphelidium* und virusartige Partikel im Cytoplasma infizierter *Scenedesmus* Zellen. Arch Microbiol 75: 209–299

Semancik JS, Vidaver AK, Van Etten JL (1973) Characterization of a segmented double-helical RNA from bacteriophage $\phi6$. J Mol Biol 78: 617–625

Shatkin AJ, Sipe JD, Loh P (1968) Separation of ten reovirus genome segments by polyacrylamide gel electrophoresis. J Virol 2: 986–991

Tikhonenko TI (1978) Viruses of fungi capable of replication in bacteria (PB viruses). In: Fraenkel-Conrat HF, Wagner RR (eds) Comprehensive virology, vol 12. Plenum Press, New York, pp 235–269

Tuveson RW, Peterson JF (1972) Virus-like particles in certain slow-growing strains of *Neurospora crassa.* Virology 47: 527–531

Tuveson RW, Sargent ML, Bozarth RF (1975) Purification of a small virus-like particle from strains of *Neurospora crassa.* Abstr Annu Meet Am Soc Microbiol, p 216

Vodkin MH, Fink GR (1973) A nucleic acid associated with a killer strain of yeast. Proc Natl Acad Sci USA 70: 1069–1072

Volkoff O, Walters T (1970) Virus-like particles in abnormal cells of *Saccharomyces carlsbergensis.* Can J Genet Cytol 12: 621–626

Wickner RB (1976) Killer of *Saccharomyces cerevisiae:* a double-stranded ribonucleic acid plasmid. Bacteriol Rev 40: 757–773

Wood HA, Bozarth RF (1972) Properties of viruslike particles of *Penicillium chrysogenum:* one double-stranded RNA molecule per particle. Virology 47: 604–609

Wood HA, Bozarth RF (1973) Heterokaryon transfer of viruslike particles associated with a cytoplasmically inherited determinant in *Ustilago maydis.* Phytopathology 63: 1019–1021

Wood HA, Bozarth RF, Mislivec PB (1971) Viruslike particles associated with an isolate of *Penicillium brevi-compactum.* Virology 44: 592–598

Wood HA, Bozarth RF, Adler J, Mackenzie DW (1974) Proteinaceous virus-like particles from an isolate of *Aspergillus flavus.* J.Virol 13: 532–534

Virion-Associated RNA Polymerases of Double-Stranded RNA Mycoviruses

K.W. BUCK

Department of Biochemistry, Imperial College of Science and Technology, London SW7 2 AZ/United Kingdom

1 Introduction

Study of the replication cycle of a virus ideally requires a system for obtaining synchronous infection of the host cells with the virus particles. Such systems have been available for many years for bacteriophages and animal viruses, and in recent years also for plant viruses, but a suitable system has not yet been developed for any double-stranded RNA (dsRNA) mycovirus. The protoplast system described by Lhoas (1971) seems to be the most promising and there are no theoretical reasons why the method should not be adapted to obtain a synchronous infection. Success in the future may depend on (1) obtaining a virus preparation with a high proportion of infective particles, (2) fractionation of the virus preparation to eliminate replicative intermediates and (3) preparation of protoplasts in the correct physiological state for uptake of a large number of virus particles per protoplast. However even in the absence of synchronous infections, useful information concerning the replication cycle of a virus can often be obtained by indirect methods, including studies in vitro of virion-associated nucleic acid polymerases, which will form the subject of the present paper.

The presence of transcriptase activity in the virion is a requirement for the infectivity of several different groups of viruses (Table 1). (1) In the case of eukaryotic DNA viruses which replicate in the cytoplasm, transcriptase is required to produce virus

Table 1. Virion-associated nucleic acid polymerases

Baltimore class[a]	Reaction catalysed	Examples of virus family	References
I	$\pm$ DNA $\rightarrow$ mRNA	Poxviridae	b
III	$\pm$ RNA $\rightarrow$ mRNA	Reoviridae	c
V	$-$ RNA $\rightarrow$ mRNA	Rhabdoviridae	d
		Orthomyxoviridae	e
		Paramyxoviridae	f
		Bunyaviridae	g
		Arenaviridae	h
VI	$+$ RNA $\rightarrow -$DNA $\rightarrow \pm$ DNA	Retroviridae	i

[a] Baltimore (1971); [b] Kates and Beeson (1970); [c] Joklik (1974); [d] Wagner (1975); [e] Skehel (1971); [f] Choppin and Compans (1975); [g] Ranki and Pettersson (1975); [h] Carter et al. (1974); [i] Baltimore (1970, 1976)

messenger RNAs (mRNAs), since host DNA-dependent RNA polymerases are confined to the nucleus. (2) With RNA viruses, in which the virion RNA is not a messenger RNA, such as dsRNA viruses and negative strand RNA viruses, transcriptase is required to produce virus mRNAs, since RNA-dependent RNA polymerases are not thought to occur in uninfected cells. (3) In the case of retroviruses virion reverse transcriptase is required to produce a DNA copy of the virion RNA and virus mRNA is then transcribed from the DNA provirus by host RNA polymerase in the nucleus. It is significant that free virion nucleic acid isolated from any of these viruses is not infective. It is also noteworthy that free virion nucleic acid is never released from the virus nucleocapsids during the infection cycle. It is likely that interaction of virion nucleic acid with the internal proteins of the nucleocapsid maintains it in a conformation suitable for recognition by the polymerase and for allowing an efficient reinitiation process in multiple rounds of transcription.

Table 2. ds-RNA mycoviruses in which virion-associated RNA polymerase activity has been identified

Virus	Reaction catalysed in vitro	References
Penicillium stoloniferum virus S	dsRNA → dsRNA	a
P. stoloniferum virus F	dsRNA → dsRNA?	b
P. chrysogenum virus	?	c
P. cyaneo-fulvum virus	?	d
Aspergillus foetidus virus S	dsRNA → ssRNA	e
A. foetidus virus F	?	f
Saccharomyces cerevisiae virus	ssRNA → dsRNA	g
	dsRNA → ssRNA	h
Allomyces arbuscula virus	dsRNA → ssRNA	i
Gaeumannomyces graminis virus	?	j
Phialophora radicicola virus 1	dsRNA → ssRNA	k
P. radicicola virus 2	?	k

[a] Buck (1975); [b] Chater and Morgan (1974); [c] Nash et al. (1973); [d] K.W. Buck and R.F. Girvan (unpublished results); [e] Ratti and Buck (1979); [f] Ratti and Buck (1975); [g] Bevan and Herring (1976); [h] Herring and Bevan (1977); [i] Khandjian and Turian (1977); [j] K.W. Buck and C.J. Rawlinson (unpublished results); [k] K.W. Buck, R. McGinty, and C.J. Rawlinson (unpublished results)

In addition to transcriptase activity virions may also contain a number of other enzyme activities required for processing the transcripts e.g., the negative strand RNA virus vesicular stomatitis virus contains the enzymes necessary to generate the 7-methyl-guanosine containing cap at the 5' terminus and the polyadenylate tail at the 3' terminus of the mRNA (Banerjee and Rhodes, 1973; Rhodes et al., 1974).

The dsRNA mycoviruses in which a virion-associated RNA polymerase has been detected in vitro are listed in Table 2. In this paper I shall discuss the nature and mechanism of formation of the products obtained in vitro with the virion RNA polymerases of two of these viruses, namely *Penicillium stoloniferum* virus S(PsV-S) and *Aspergillus foetidus* virus S (AfV-S) and the implications of these results for the replication cycles in vivo of these viruses. These two enzyme activities will then be compared briefly with those found in virions of other dsRNA mycoviruses and of dsRNA viruses of other host taxa.

2 The Virion RNA Polymerase of Penicillium stoloniferum Virus S

PsV-S has a genome of two segments of dsRNA with molecular weights of 0.94×10^6 and 1.11×10^6 (Bozarth et al., 1971; Buck and Kempson-Jones, 1973). Each of these is probably monocistronic, one coding for the capsid polypeptide, MW 42,500 and the other for the single chain RNA polymerase, MW 55,500 (Buck and Kempson-Jones, 1974). RNA polymerase activity in virions of PsV-S was first demonstrated by Lapierre et al. (1971) and later Chater and Morgan (1974) showed that the products of reaction were two dsRNA molecules, with the same molecular weights as genome dsRNAs, and that these products remained within the virus particles.

In order to examine this RNA polymerase reaction in more detail Buck (1975) examined the distribution of activity among the different particle types of the PsV-S multicomponent system. In order to produce sufficient material for these studies virus was purified from mycelium obtained from growth of *P. stoloniferum* in a 60 litre fermenter. Coprecipitation of the virus with yeast RNA at pH 4.0, used previously for concentrating this virus (Banks et al., 1971) could not be used in these studies, because the RNA polymerase was inactivated under these conditions. However precipitation with 10% w/v polyethylene glycol 6000 and 0.5 M-sodium chloride was found to be a useful

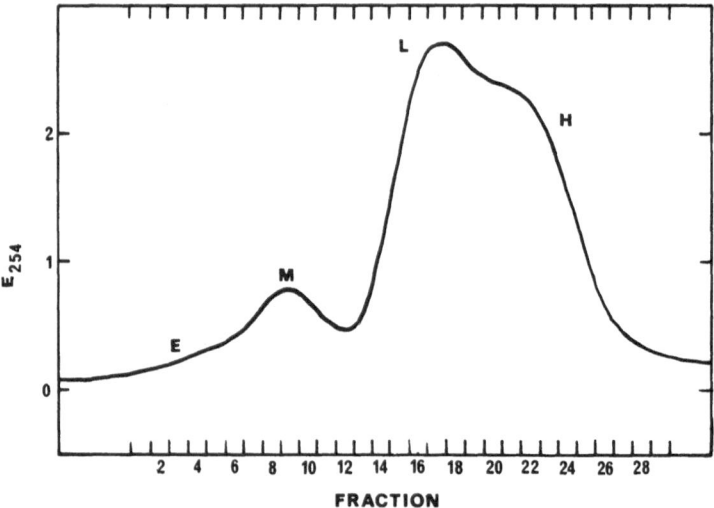

Fig. 1. Sucrose density gradient centrifugation of PsV-S. A preparation of PsV-S (50 ml, E_{260} 10) was loaded onto a 10 to 45% w/w sucrose gradient (500 ml) in an M.S.E. BIV zonal rotor and centrifuged at 45,000 rev/min for 4 h. After separation the gradient was pumped through an ISCO flow cell and 10 ml fractions were collected

alternative and resulted in virus preparations with high polymerase activity. Virus was separated by velocity centrifugation in a sucrose density gradient in an M.S.E. B XIV zonal rotor. The extinction profile is shown in Figure 1. The separation into four bands, designated E, M, L, and H, obtained with the 500 ml zonal rotor was comparable to that obtained previously on a small scale with the 5 ml Beckman SW50 rotor (Buck and Kempson-Jones, 1973). However caesium chloride gradients could not be used for virus separation because of the resultant loss of polymerase activity.

Band E was found to contain empty capsids, whereas bands M, L, and H contained particles with RNA. Figure 2 shows the results of the analysis of RNA prepared from selected fractions by electrophoresis in polyacrylamide gels containing 8 M-urea. Frac-

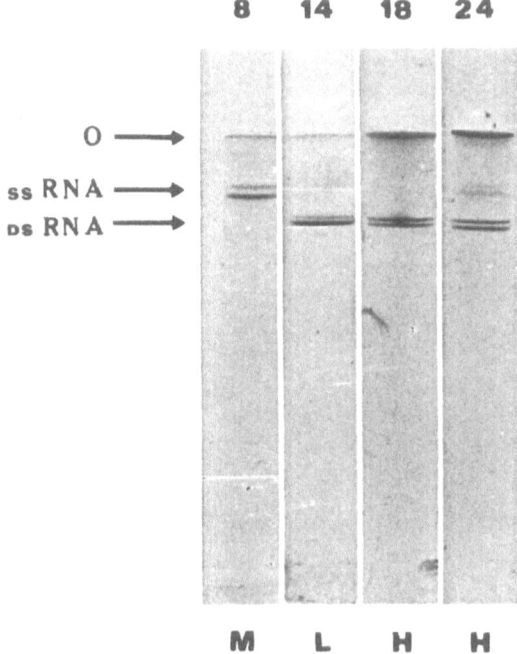

Fig. 2. Analysis of RNA prepared from sucrose gradient fractions 8, 14, 18, and 24 by electrophoresis in 4% polyacrylamide gels containing 8 M-urea according to the method described by Buck and Ratti (1977)

tion 8 (M particles) shows two bands of single-stranded RNA (ssRNA), which have molecular weights of exactly one half of the virion dsRNA components. The ssRNA molecules are the virus mRNAs. If these two RNA molecules are used to programme a wheat germ system for in vitro protein synthesis, virus capsid protein is the principal product (K.W.Buck, unpublished results). Fraction 14 (L particles) gave only two bands of the genome dsRNA, while fraction 18 (L particles plus low density H particles) gave the two bands of dsRNA, plus material at the top of the gel, which was sensitive to ribonuclease A in 0.3 M-sodium chloride and therefore contained ssRNA. Fraction 24 (high density H particles) gave a similar pattern of bands to fraction 18, but in addition showed two bands of ssRNA with the same mobilities as the ssRNAs from M particles.

The capsids of M, L, and H particles are identical and are composed of 120 molecules of the polypeptide, MW 42,500; in addition all the particles contain one molecule of the single chain RNA polymerase, MW 55,500. However only H particles were found to be active in the in vitro RNA polymerase assay system. After reaction the RNA on top of the gel could no longer be detected and is therefore assumed to consist of intermediates of the reaction.

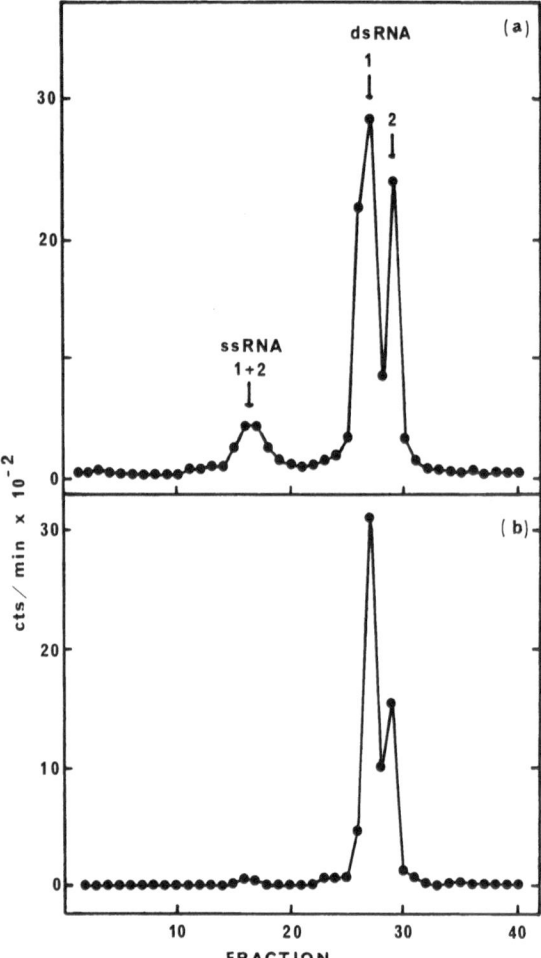

Fig. 3. Analysis of the RNA product from polymerase reactions with a fraction 18 and b fraction 24 carried out in the presence of ^{3}H-UTP. Electrophoresis was carried out in 4% polyacrylamide gels containing 8 M-urea according to the method described by Buck and Ratti (1977). 0.5 mm slices were cut, starting at the top of the gel

Analysis by electrophoresis in urea gels of the products of RNA polymerase reactions with fractions 18 and 24 carried out in the presence of ^{3}H-UTP is shown in Figure 3. The major labelled products are dsRNAs 1 and 2, confirming the results of Chater and Morgan (1974), but from the low density H particles (fraction 18), small amounts of labelled ssRNAs 1 and 2 were also detected. Examination of the intact virus particles after reaction by preparative isopycnic caesium chloride density gradient centrifugation showed all the label to be in dense particles (designated product or P particles), derived from H particles, while M and L particles were unlabelled, and unchanged in amount or density.

Since H particles contained intermediates of the RNA polymerase reaction and were heterogeneous in density (Fig. 4a), it was concluded that the RNA intermediates were probably also heterogeneous, ranging from those in which reactions had just been initiated to those near completion. In order to study the complete reaction and to obtain a reasonable degree of synchrony it was necessary to select the low density H particles, since these were the particles in which the reaction had just been initiated and in which the maximum amount of RNA synthesis per H particle should be achieved.

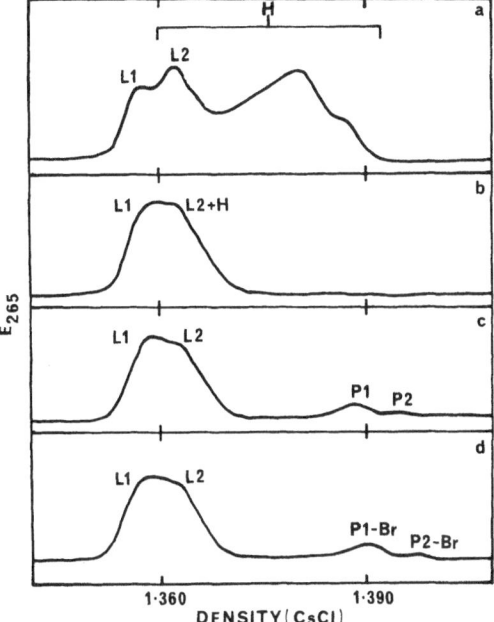

Fig. 4a-d. Analytical caesium chloride equilibrium density gradient centrifugation: UV scanner traces. Samples were adjusted to a density of 1.36 g/ml by addition of solid caesium chloride and centrifuged at 34,000 rev/min for 24 h in an An-F rotor in a Beckman Model E analytical ultracentrifuge. a Unseparated PsV-S; b fraction 16 (Fig. 1) before reaction; c fraction 16 (Fig. 1) after reaction in the presence of UTP; d fraction 16 (Fig. 1) after reaction in the presence of 5-bromo-UTP

Figure 4b shows a caesium chloride equilibrium gradient profile obtained from fraction 15 (Fig. 1), which contains mainly L particles, together with a small quantity of low density H particles. After reaction the L particles are unchanged, while the H particles are converted to the more dense P particles (Fig. 4c). Measurement of the specific activity of the dsRNA in P particles, isolated after a reaction containing ^{3}H-UTP, showed that one complete molecule of dsRNA per particle was newly synthesised. Since the particles started with one molecule of dsRNA, P particles must contain two molecules of dsRNA and the overall reaction is therefore a complete replication of dsRNA.

The next question to answer was whether the reaction occurred conservatively or semi-conservatively. This problem was tackled by using density labelling in an experiment similar in principle to that used 20 years ago for establishing semi-conservative replication of DNA in *Escherichia coli* (Meselson and Stahl, 1958). Trial experiments using incorporation of ^{3}H-ATP into acid-insoluble product established that 5-bromo-UTP could efficiently replace UTP in the replication reaction and caesium chloride equilibrium density gradient analysis of the product showed the production of P particles, as with UTP (Fig. 4d), albeit slightly more dense.

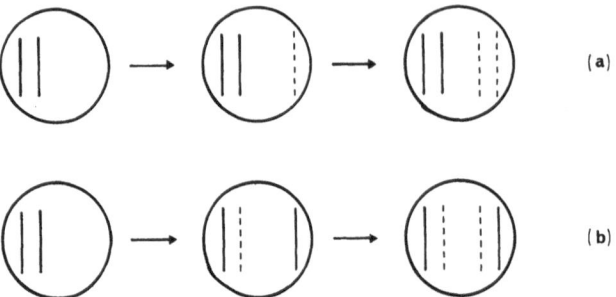

Fig. 5. Diagrammatic representation of the possibilities of **a** conservative and **b** semi-conservative replication in particles of PsV-S. *Unbroken lines* represent virion RNA before reaction. *Broken lines* represent newly synthesised RNA after in vitro RNA polymerase reaction and do not imply fragmentation of the product. Capsids are shown in *circles*

RNA was isolated from these bromo-P particles and centrifuged to equilibrium in a caesium sulphate density gradient in the analytical ultracentrifuge. Figure 5 shows the theoretical possibilities of the products of (1) conservative and (2) semi-conservative replication. If replication is conservative, P particles should contain one molecule of dense dsRNA labelled in both strands (Br U:Br U RNA) and one molecule of dsRNA unlabelled (U:U RNA) and hence of original density. On the other hand if replication is semi-conservative, the product will consist of two molecules of dsRNA each labelled in one strand (Br U:U RNA) and no unlabelled dsRNA of original density will remain. The results showed that only one band of dsRNA with a buoyant density of 1.647 g/ml was obtained from the bromo-P particles and none of the original (U:U) dsRNA, buoyant density 1.606 g/ml, could be detected. Replication of dsRNA in vitro in virions of PsV-S therefore takes place by a semi-conservative mechanism (Buck, 1978).

A consequence of the semi-conservative mechanism is that the RNA polymerase must use a dsRNA template for the first stage of the reaction and the displaced molecule of ssRNA for the second stage of the reaction. Although in the in vitro system the polymerase appears to be unable to initiate RNA synthesis on a ssRNA template in M particles or on a dsRNA template in L particles, it must terminate the reaction on the dsRNA template and reinitiate the reaction on the ssRNA template in the second stage of the replication reaction in H particles. One possibility that overcomes the problem of re-initiation is that the 3'end of the ssRNA template could loop round to form a hairpin structure and act as a primer for further RNA synthesis. The product of the reaction would then be a snap-back RNA (Fig. 6); molecules of this type have been isolated from defective particles of vesicular stomatitis virus (Perrault and Leavitt, 1978).

In order to test this hypothesis the ^{3}H-labelled dsRNA product of a replication reaction was denatured by treatment with 60% w/v dimethyl sulphoxide and renaturation was prevented by heating with glyoxal essentially as described by McMaster and Car-

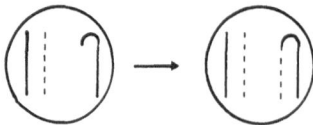

Fig. 6. Diagrammatic representation of the possible formation of a snap-back RNA in the second stage of the semi-conservative replication of dsRNA in particles of PsV-S. Notation as in Figure 5

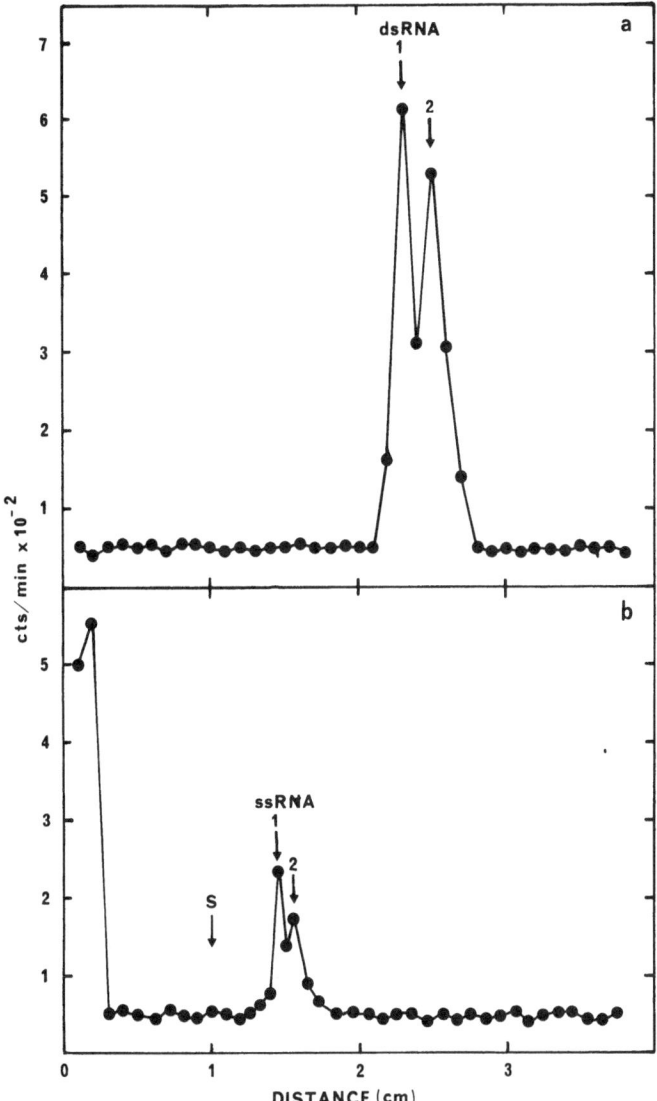

Fig. 7. Analysis of the product of PsV-S RNA polymerase reaction by electrophoresis in 4% poly-acrylamide gels containing 8 M-urea by the method described by Buck and Ratti (1977): **a** RNA isolated from an RNA polymerase reaction in the presence of ^{3}H-UTP with unfractionated PsV-S; **b** RNA in **a**, after denaturation in dimethyl sulphoxide and glyoxylation. S marks the position of 23 S *Escherichia coli* ribosomal RNA (MW 1.05 × 10^6)

michael (1977). The denatured RNA was then analysed by electrophoresis in polyacryl-amide gels containing 8 M-urea. If snap-back RNAs are formed they should denature to form ssRNAs of molecular weights equal to that of the original dsRNAs 1 and 2 (i.e. about 1 × 10^6), whereas if the dsRNA products are composed of two individual strands they should denature to give ssRNA molecules with the same molecular weights as ssRNAs 1 and 2 (i.e., about 0.5 × 10^6). The results in Figure 7 show gels of the product

RNA before and after denaturation. After denaturation two bands with the same mobilities as ssRNAs 1 and 2 were formed, but no band in the region of a ssRNA of MW
1×10^6 could be detected. The results do not therefore support the snap-back mechanism, but are not completely unequivocal in view of the considerable amount of radioactivity remaining on top of the gel. However further studies on the kinetics of reasso-
ciation of the denatured product dsRNA have indicated the absence of significant
amounts of snap-back RNA molecules among the products of the PsV-S RNA replication reaction (K.W. Buck and R. Empson, unpublished results). These experiments do
not exclude the possibility that a snap-back molecule is formed and then cleaved in
the loop by a nuclease. Asymmetric cleavage in such a loop could generate a short single-
stranded stretch at the 5' terminus of one strand, such as has been suggested for *P. chrysogenum* dsRNA (Yazaki and Miura, 1977) and dsRNA3 of *P. stoloniferum* virus F
(Szekely and Loviny, 1975).

Finally it has to be considered whether the symmetrical replication reaction in virions of PsV-S occurs in vivo. It is clear that the first stage of the reaction does occur
in vivo, since there is good evidence that the more dense H particles contain one molecule of dsRNA and one molecule of ssRNA transcript. However there is no evidence at
present for P particles or intermediates between the halfway and complete reaction i.e.
partially double-stranded molecules in PsV-S preparations. Therefore the possibility
should be considered that the in vitro replication reaction could occur because the
ssRNA transcript is not ejected from the particles under in vitro conditions, but remains
to act as a template for further polymerase reaction.

Two possible pathways for the replication of PsV-S dsRNA in vivo are represented
diagrammatically in Figure 8. The in vitro studies show that the synchronous pathway

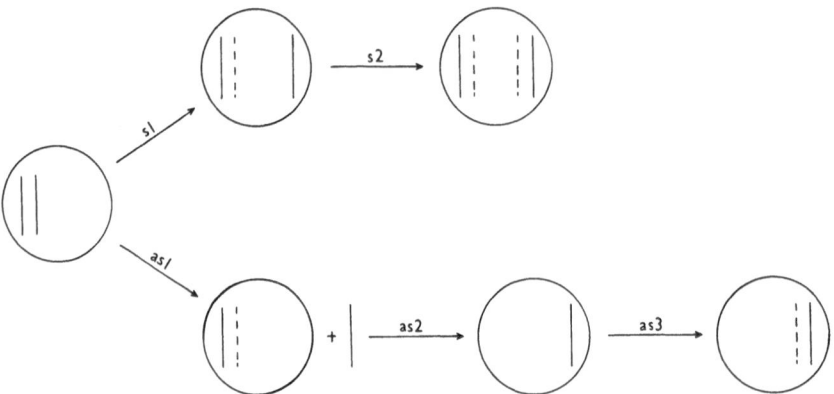

Fig. 8. Possible pathways for the replication of PsV-S dsRNA in vivo. (s1, s2) synchronous replication. Strand displacement is followed immediately by formation of dsRNA on the displaced strand
as template *in the same particle;* (as1, as2, as3) asynchronous replication. The displaced strand
is released from the parental particle and is later encapsidated to form *progeny nascent virions* in
which it serves as a template for dsRNA synthesis. Notation as in Figure 5

(s1 and s2) can occur in vitro, but do not prove that s2 occurs in vivo. Recently we
have shown that transcription of dsRNA and release of ssRNA from virions (as1) occurs

in vitro in a small fraction of virus particles (see also Fig. 3a), and the isolation from infected cells of M particles, which contain ssRNA and the RNA polymerase polypeptide, suggests that the asynchronous pathway (as1 → 3) probably occurs in vivo. However whether replication occurs in vivo by asynchronous or synchronous pathways (or both), the mechanism is semi-conservative.

3 The Virion RNA Polymerase of Aspergillus foetidus Virus S

AfV-S differs from PsV-S in that it has a much larger genome, consisting of three components, dsRNA1 (MW 4.0×10^6), dsRNA2 (MW 2.6×10^6) and dsRNA3 (MW 0.27×10^6) (Buck and Ratti, 1977). Comparison of the molecular weights of dsRNAs 1 and 2 with the molecular weight of the single capsid polypeptide (83,000) (Buck and Ratti, 1975) suggests that these two RNAs are at least dicistronic. AfV-S has four basic virions (RNA components encapsidated in parentheses): S1a (dsRNA2); S1b (dsRNA2 + dsRNA3); S2a (dsRNA1); S2b (dsRNA1 + dsRNA3). In addition virus preparations contain small numbers of particles (S2x, S3 and S4) which are believed to contain intermediates or products of RNA transcription or replication. Virions have RNA polymerase activity (Ratti and Buck, 1975), but no minor polypeptide of the virion has as yet been equated with this activity.

The virion RNA polymerase of AfV-S has been shown to be a transcriptase (Ratti and Buck, 1979). The major product of reaction is full length ssRNA transcripts of one of the strands of dsRNA2 and these transcripts are released from the virions. Re-initiation of RNA synthesis occurs in the in vitro system and the reaction continues for up to 48 h after which 6 to 8 copies of ssRNA are produced for each molecule of dsRNA2 present. Most of the transcription of dsRNA2 occurs in S1a particles. Small numbers of complete transcripts of dsRNA1 were also formed, but no transcription of dsRNA3 could be detected.

During transcription in the presence of tritiated nucleoside triphosphate (nTP) substrates, dsRNA2 becomes labelled within the virions, the amount of newly synthesised RNA being equivalent to one strand in 40% to 80% of dsRNA2 molecules. Most of this activity is also associated with S1a particles. The amount of RNA synthesis is too great to be accounted for by "filling up" of possible ssRNA tails on dsRNA2 by a ss → ds RNA polymerase, since dsRNA2 isolated from S1a particles contains no ssRNA detectable by a change in electrophoretic mobility in polyacrylamide gel after treatment with ribonuclease A in 0.3 M-NaCl (Ratti and Buck, 1972). Two other possibilities were considered to account for the labelling of dsRNA2: (1) replication of dsRNA within S1a particles could occur by analogy with the PsV-S system; (2) the AfV-S transcription reaction could occur by displacement of one strand of dsRNA by the strand being newly synthesised, so that at the end of the first round of transcription, the displaced ssRNA strand is released from the particle and the dsRNA remaining within the virion contains one conserved and one newly synthesised strand i.e., the reaction is semi-conservative with respect to dsRNA. Several lines of evidence (Ratti and Buck, 1978) support the semi-conservative transcription mechanism, illustrated diagrammatically in Figure 9.

1. Competition hybridisation studies have shown that only one strand of dsRNA2 becomes labelled and that this has the same sequence as the ssRNA2 transcripts produced.

2. Labelling of dsRNA2 reaches a maximum after 4 h. However if the polymerase reaction is carried out for 4 h with unlabelled nTPs and then tritiated nTPs are added, label is still incorporated into dsRNA2 at a rate consistent with the rate of formation of ssRNA2 transcripts.

3. If reactions are carried out for a short period (pulse) with labelled nTPs and then a large excess of unlabelled nTPs are added, most of the label incorporated into dsRNA2 during the pulse can be chased into ssRNA2 by subsequent reaction. Replicative intermediates formed during the pulse consist of dsRNA2 molecules with ssRNA tails, as shown in Figure 9.

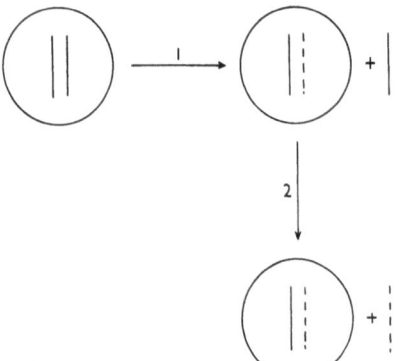

Fig. 9. Transcription by semi-conservative strand displacement in particles of AfV-S. Notation as in Figure 5. Transcripts released after round 1 are unlabelled, and those released after round 2 are labelled

4. In reactions with S1a particles and labelled nTPs the first ssRNA2 transcripts to be released were unlabelled, as expected for the first round of transcription by semi-conservative strand displacement (Fig. 9).

The above experiments do not eliminate the possibility that some replication of dsRNA2 occurs in a small proportion of S1a particles. AfV-S preparations contain small numbers of S4 particles, which contain two molecules of dsRNA2 (Buck and Ratti, 1975), suggesting that such a reaction could occur in vivo. (Direct encapsidation of two molecules of dsRNA2 in one particle is, of course, an alternative possibility). No particles containing only ssRNA or intermediates of a ss → ds RNA polymerase reaction have been detected in preparations of AfV-S. Hence, as with PsV-S, we can deduce from the in vitro RNA polymerase experiments that transcription and replication of dsRNA2 in AfV-S occurs semi-conservatively in vivo, but the available evidence does not allow us to decide whether dsRNA replication in vivo occurs by a synchronous or asynchronous pathway (Fig. 8).

4 Virion RNA Polymerases of Other dsRNA Mycoviruses

Only those cases where the products of polymerase reaction have been characterised will be discussed.

4.1 Saccharomyces cerevisiae Virus

Virus particles obtained from sensitive strains of *S. cerevisiae* contain one molecule of L (or P1) dsRNA (MW ca. 3×10^6) and a capsid composed principally of polypeptide subunits, MW 75,000 (Wickner, 1976). Virus particles obtained from either log or stationary phase cells exhibit RNA polymerase activity in vitro; as yet no minor virion polypeptide has been equated with this activity. L (P1) dsRNA has been shown by in vitro protein synthesis to code for the capsid polypeptide (Hopper et al., 1977). As with dsRNAs 1 and 2 of AfV-S it is likely that L (P1) dsRNA is at least dicistronic and it probably also encodes the virus RNA polymerase.

In virus preparations isolated from log phase cells RNA polymerase activity lay to the slower sedimenting side of the main virus peak in sucrose density gradient centrifugation and was shown to consist of a ss → ds RNA polymerase (Bevan and Herring, 1976). Examination of RNA from active particles by electron microscopy revealed dsRNA molecules, considerably shorter than L (P1) dsRNA, with long ssRNA tails. In virus preparations isolated from stationary phase cells the activity coincided with the main virus peak and the principal reaction product was complete ssRNA transcripts of L (P1) dsRNA (Herring and Bevan, 1977). Yeast virus therefore has both of the RNA polymerase activities required for the asynchronous pathway of dsRNA replication (Fig. 8). L (P1) dsRNA became labelled to a small extent during the transcription reaction, but the available data do not allow a conclusion to be reached as to whether this was due to (a) ss → ds RNA polymerase activity, (b) ds → dsRNA polymerase activity, (c) transcription by semi-conservative strand displacement.

4.2 Allomyces arbuscula Virus

The major products of the virion RNA polymerase reaction in vitro of *A. arbuscula* virus were found to be dsRNAs with the same molecular weights as those of the three genome dsRNAs i.e., 2.4×10^6, 1.3×10^6 and 1.0×10^6; newly synthesised dsRNA remained within virions (Khandjian and Turian, 1977). Thirty five percent of the label in reactions containing tritiated substrates was incorporated into ssRNA molecules (MW 0.5×10^6), which were released from the particles and are probably transcripts of the smallest dsRNA molecule. The mechanism of incorporation of label into dsRNA is not known.

4.3 Phialophora radicicola Virus 1

RNA polymerase activity has been detected in virions of a dsRNA virus isolated from *P. radicicola* var. *radicicola* (K.W. Buck, R. McGinty and C.J. Rawlinson, unpublished results). The products of reaction were ssRNA transcripts, which were released from the particles, and virion dsRNA became labelled in reactions in the presence of tritiated nTPs.

5 Comments

The studies described above allow the conclusion that transcription and replication of dsRNA in PsV-S and AfV-S occur by a semi-conservative mechanism, but it is too early to speculate whether this mechanism will prove to be common to dsRNA mycoviruses in general. Yeast virus can clearly replicate its dsRNA by the asymmetrical pathway (Fig. 8). The studies with PsV-S suggest that some dsRNA mycoviruses may be able to replicate their dsRNA by either synchronous or asynchronous pathways; the choice could perhaps be governed by the physiological state of the host (e.g., log/stationary cultures) or by early/late switches in the virus replication cycle. DsRNA mycoviruses are not a homogeneous group; dsRNA molecular weights range from 0.27×10^6 to 6.3×10^6, polypeptide molecular weights range from 42,000 to 130,000 and numbers of RNA components vary from 1 up to 5 in different viruses. It would therefore be surprising if there were not minor, or even major, differences in the replication strategies of individual viruses.

6 Virion RNA Polymerases of dsRNA Viruses of Other Host Taxa

The properties of virion RNA polymerases of dsRNA viruses of other host taxa will be described briefly in order to facilitate comparisons with those of dsRNA mycoviruses.

6.1 Reoviridae

This is a large virus family, comprising several genera, and including viruses of vertebrates, invertebrates and plants. Members have double-shelled icosahedral particles with genomes of 10 to 12 molecules of dsRNA all enclosed together in single virions (Joklik, 1974). Virions, or cores derived from virions, have transcriptase activity and the ssRNA copies (mRNAs) of all the genome dsRNA molecules which are monocistronic, are released from the particles. Synthesis of dsRNA occurs in separate subviral particles, containing one copy of each of the virus mRNAs, by a ss → dsRNA polymerase. Replication of dsRNA is therefore *asynchronous*. However transcription and replication of dsRNA occur *conservatively*.

6.2 Cystoviridae

The only known member of this family, at present, is bacteriophage $\phi6$ from *Pseudomonas phaseolicola* (Semancik et al., 1973). It is a lipid-enveloped virus with an icosahedral nucleocapsid and a genome comprising three dsRNA components (MW 4.9×10^6, 2.9×10^6, and 2.0×10^6) which are all polycistronic. Replication of $\phi6$ dsRNA in vivo occurs *semi-conservatively* and *asynchronously* (Coplin et al., 1975). The phage carries an RNA polymerase, which in vitro catalyses the formation of ssRNA transcripts of the two smaller dsRNA components (Partridge et al., 1978). DsRNA also becomes labelled during the reactions (Van Etten et al., 1973; Liljestrom et al., 1978) and it seems likely that $\phi6$ polymerase catalyses semi-conservative transcription of dsRNA as shown for AfV-S.

6.3 Animal Viruses with Bisegmented dsRNA Genomes

RNA polymerase activity has been demonstrated in virions of *Drosophila* X virus, a member of a new group of viruses with genomes consisting of two segments of dsRNA (MW 2.3 to 2.6 × 10^6), which includes viruses of vertebrates and invertebrates (Dobos et al., 1978; Barnard and Petitjean, 1978). DsRNA was synthesised during the reaction and it was suggested that the enzyme could be a dsRNA replicase similar to that of PsV-S (Barnard, 1978).

Summary

Virion-associated nucleic acid polymerases are requirements for the infectivity of viruses which do not have a genome with a messenger RNA function, and for which cellular enzymes are not able, or are not available, to transcribe directly the virion genome to produce messenger RNA. These include DNA viruses which replicate in the cytoplasm, double-stranded RNA viruses, negative strand RNA viruses and RNA tumour viruses. Virion-associated RNA polymerases have now been detected in several double-stranded RNA mycoviruses. In *Penicillium stoloniferum* virus S the virion RNA polymerase catalyses in vitro semi-conservative replication of double-stranded RNA, whereas that of *Aspergillus foetidus* virus S catalyses transcription by semi-conservative strand displacement. Recent investigations with these two RNA polymerases are described and the implications with regard to the replication cycles in vivo of double-stranded RNA mycoviruses and double-stranded RNA viruses of other host taxa are discussed.

References

Baltimore D (1970) RNA-dependent DNA polymerase in virions of RNA tumour viruses. Nature (London) 226: 1209–1211
Baltimore D (1971) Expression of animal virus genomes. Bacteriol Rev 35: 235–241
Baltimore D (1976) Viruses, polymerases and cancer. Science 192: 632–636
Banerjee AK, Rhodes DP (1973) In vitro synthesis of RNA that contains polyadenylate by virion-associated RNA polymerase of vesicular stomatitis virus. Proc Natl Acad Sci USA 70: 3566–3570
Banks GT, Buck KW, Fleming A (1971) The isolation of viruses and viral ribonucleic acid from filamentous fungi on a pilot plant scale. Chem Eng 251: 259–261
Barnard JP (1978) An RNA polymerase activity associated with *Drosophila* X virus. Abstr 4th Int Congr Virol 43: 574
Barnard JP, Petitjean AM (1978) In vitro synthesis of double-stranded RNA by *Drosophila* X virus purified virions. Biochem Biophys Res Commun 83: 763–770
Bevan EA, Herring AJ (1976) The killer character in yeast: preliminary studies of virus-like particle replication, In: Bandolow W, Schweyen RJ, Thomas DY, Wolfe K, Kauderwitz F (eds) Genetics, biogenesis and bioenergetics of mitochondria. Walter de Gruyter, Berlin, pp 153–163
Bozarth RF, Wood HA, Mandelbrot A (1971) The *Penicillium stoloniferum* virus complex: two similar double-stranded RNA virus-like particles in a single cell. Virology 45: 516–523
Buck KW (1975) Replication of double-stranded RNA in particles of *Penicillium stoloniferum* virus S. Nucleic Acids Res 2: 1889–1902
Buck KW (1978) Semi-conservative replication of double-stranded RNA by a virion-associated RNA polymerase. Biochem Biophys Res Commun 84: 639–645
Buck KW, Kempson-Jones GF (1973) Biophysical properties of *Penicillium stoloniferum* virus S. J Gen Virol 18: 223–235

Buck KW, Kempson-Jones GF (1974) Capsid polypeptides of two viruses isolated from *Penicillium stoloniferum.* J Gen Virol 22: 441–445

Buck KW, Ratti G (1975) Biophysical and biochemical properties of two viruses isolated from *Aspergillus foetidus.* J Gen Virol 27: 211–224

Buck KW, Ratti G (1977) Molecular weight of double-stranded RNA: a re-examination of *Aspergillus foetidus* virus S RNA components. J Gen Virol 37: 215-219

Carter MF, Biswal N, Rawls WE (1974) Polymerase activity of Pichinde virus. J Virol 13: 577–583

Chater KF, Morgan DH (1974) Ribonucleic acid synthesis by isolated viruses of *Penicillium stoloniferum.* J Gen Virol 24: 307–317

Choppin PW, Compans RW (1975) Reproduction of paramyxoviruses. In: Fraenkel-Conrat H, Wagner RR (eds) Comprehensive virology, vol IV. Plenum Press, New York, pp 95–178

Coplin DL, Van Etten JL, Koshi RD, Vidaver AK (1975) Intermediates in the biosynthesis of double-stranded ribonucleic acids of bacteriophage φ6. Proc Natl Acad Sci USA 72: 849–853

Dobos P, Hallett R, Kells DTC, Hill BJ, Becht H, Teninges D (1978) Biophysical and biochemical studies of five animal viruses with bi-segmented dsRNA genomes. Abstr 4th Int Congr Virol 23: 335

Herring AJ, Bevan EA (1977) Yeast virus-like particles possess a capsid-associated single-stranded RNA polymerase. Nature (London) 268: 464–466

Hopper JE, Bostian KA, Rowe LB, Tipper DJ (1977) Translation of the L-species dsRNA genome of the killer-associated virus-like particles of *Saccharomyces cerevisiae.* J Biol Chem 252: 9010–9017

Joklik WK (1974) Reproduction of reoviridae. In Fraenkel-Conrat H, Wagner RR (eds) Comprehensive virology, vol IV. Plenum Press, New York, pp 231–334

Kates J, Beeson J (1970) Ribonucleic acid synthesis in vaccinia virus I. The mechanism of synthesis and release of RNA in vaccinia cores. J Mol Biol 50: 1–18

Khandjian EW, Turian G (1977) In vitro RNA synthesis by double-stranded RNA mycovirus from *Allomyces arbuscula.* FEMS Microbiol Lett 2: 121–124

Lapierre H, Astier-Manifacier S, Cornuet P (1971) Activité RNA polymérase associée aux préparations purifiées de virus du *Penicillium stoloniferum.* C R Acad Sci (Paris) Ser D 273: 992–994

Lhoas P (1971) Infection of protoplasts from *Penicillium stoloniferum* with double-stranded RNA viruses. J Gen Virol 13: 365–367

Liljestrom P, Ranki M, Soderlund H, Bamford D (1978) In vitro transcription of bacteriophage φ6 double-stranded RNA. Abstr 4th Int Congr Virol 46: 593

McMaster GK, Carmichael GG (1977) Analysis of single- and double-stranded nucleic acids on polyacrylamide and agarose gels by using glyoxal and acridine orange. Proc. Natl. Acad. Sci USA 74: 4835–4838

Meselson M, Stahl FW (1958) The replication of DNA in *Escherichia coli.* Proc Natl Acad Sci USA 44: 671–682

Nash CH, Douthart RJ, Ellis LF, Van Frank RM, Burnett JP, Lemke PA (1973) On the mycophage of *Penicillium chrysogenum.* Can J Microbiol 19: 97–103

Partridge JE, Vidaver AK, Van Etten JL (1978) In vitro transcription of bacteriophage φ6 double-stranded RNA. Abstr Annu Meet Am Soc Microbiol 553: 221

Perrault J, Leavitt RW (1978) Characterisation of snap-back RNAs in vesicular stomatitis defective interfering virus particles. J Gen Virol 38: 21–34

Ranki M, Pettersson RF (1975) Uukuniemi virus contains an RNA polymerase. J Virol 16: 1420–1425

Ratti G, Buck KW (1972) Virus particles in *Aspergillus foetidus:* a multicomponent system. J Gen Virol 14: 165–175

Ratti G, Buck KW (1975) RNA polymerase activity in double-stranded ribonucleic acid virus particles from *Aspergillus foetidus.* Biochem Biophys Res Commun 66: 706–711

Ratti G, Buck KW (1978) Semi-conservative transcription in particles of a double-stranded RNA mycovirus. Nucleic Acids Res 5: 3843–3854

Ratti G, Buck KW (1979) Transcription of double-stranded RNA in virions of *Aspergillus foetidus* virus S. J Gen Virol 42: 59–72

Rhodes DP, Moyer SA, Banerjee AK (1974) In vitro synthesis of methylated messenger RNA by the virion-associated RNA polymerase of vesicular stomatitis virus. Cell 3: 327–333

Semancik JS, Vidaver AK, Van Etten JL (1973) Characterisation of a segmented double-helical RNA from bacteriophage φ6. J Mol Biol 78: 617–625

Skehel JJ (1971) RNA-dependent RNA polymerase activity of the influenza virus. Virology 45: 793–796

Szekely M, Loviny T (1975) 5'-terminal phosphorylation and secondary structure of double-stranded RNA from a fungal virus. J Mol Biol 93: 79–87

Van Etten JL, Vidaver AK, Koski RK, Semancik JS (1973) RNA polymerase activity associated with bacteriophage φ6. J Virol 12: 464–471

Wagner RR (1975) Reproduction of rhabdoviruses. In: Fraenkel-Conrat H, Wagner RR (eds) Comprehensive virology, vol IV. Plenum Press, New York, pp 1–94

Wickner RB (1976) Killer of Saccharomyces cerevisiae: a double-stranded ribonucleic acid plasmid. Bacteriol Rev 40: 757–773

Yazaki K, Miura KI (1977) Terminal structure involving a single-stranded stretch in double-stranded RNA from Penicillium chrysogenum virus. Virology 82: 14–24

Some Morphological Changes in Fungi Induced by Fungal Viruses

J. ALBOUY

Station de Pathologie Végétale, I.N.R.A. Route de Saint Clément, Beaucouze, 49000 Angers/France

1 Introduction

This paper reports observations of electron microscopic studies on some fungal viruses. Information has been obtained about the intracellular appearance of these viruses in the host and the response of infected cells: pathological lysis, typical inclusions, intracytoplasmic location and distinctive morphological structures.

There are a number of reports on the ultrastructure of mycovirus-infected tissues, many showing or confirming the presence of the virus particles in old cells or intracytoplasmic distribution. Some mycoviruses are known only as a result of ultrastructural observations.

However, the reports of pathogenic effects and morphogenesis of the particles are circumstantial observations. Since samples contained heterogeneous populations of cells infected at different times, the various steps of infection and the sequence of cytological changes are difficult or impossible to determine. Without experimental conditions suitable for infecting the hosts and making comparisons of healthy and virus infected hyphae of a similar age, it is extremely dangerous to assume that some observed changes are attributed only to infection. These modifications could be due to cell lysis with the disappearance of organelles which could be confused with normal autolysis of old cells. Other modifications such as filamentous inclusions and elaboration of membrane development are morphological structures related to infection.

In most cases we can describe only advanced stages of the infection, while the earliest and most short-lived events are not included.

2 Materials and Methods

For ultrathin sections the mycelia, obtained from liquid or agar cultures or sporophores were fixed for 3 h in 6% buffered glutaraldehyde (pH 7.0), then postfixed for 1 h in buffered osmium tetroxide. After prestaining in uranyl acetate (0.5%), dehydration with a graded ethanol series and propylene oxide, specimens were embedded in Araldite. Sections cut with an ultramicrotome LKB using a glass knife, were stained with uranyl acetate and lead citrate, and were observed in Hitachi HS 8 electron microscope. Samples were taken between 4 and 12 days of cultivation. Healthy tissue was examined in the same way.

3 Results

3.1 Ultrastructure of Some Mycovirus-Infected Fungi

3.1.1 Sclerotium cepivorum

The two types of virus particle (30 nm and 45 nm), mostly seen in mycelial cells from old cultures, were found when degeneration of cellular organelles and disappearance of cytoplasm had occurred (Lapierre et al., 1971; Albouy and Lapierre, 1971). Numerous membranous structures were observed in infected cells. Virus particles were enclosed within double membranes. These peculiar formations seemed to be organs for virus conservation and accumulation since they were liberated into the medium when cell lysis occurred, but preventing their loss of infectivity.

3.1.2 Ophiobolus graminis

In young cells it was very difficult to discern virus particles from the ribosome dense cytoplasm. So far no cellular aberration has been observed in infected cells. Although particle numbers were low, small aggregates of 30 nm particles were readily detected in aged hyphae.

3.1.3 Mycogone perniciosa

Albouy and Lapierre (1972) described features of virus-infected mycelial cells of *Mycogone perniciosa* strains MP_1 and MP_2 reported by Lapierre et al. (1972).

Attempts to find the sequence of cytological changes caused by infection suggest that the virus reaches a high titer in host cells and causes cellular aberrations.

Sequences of Infection. In a noninfected hypha the cytoplasm is dense, rich in ribosomes with few small vacuoles. Electron-dense mitochondria and lipid bodies were scattered throughout the cell.

In infected hyphae there were certain localized areas in the cytoplasm where the development of medium electron-dense regions of granular appearance was observed. It was also noted that dense areas of inclusions were formed by tightly packed particles embedded in a granular matrix. They presented some similarities with inclusion bodies of the Caulimoviruses (Lawson and Hearon, 1977). It is suggested that the matrix could be considered a viroplasmic center. These inclusions developed into large spherical masses filling the cell and various degrees of disappearance of cytoplasm and organelles were noted (Fig. 1).

At more advanced stages it was noted that these dense cells were inflated due to an accumulation of virus particles in high concentrations (Fig. 2).

For the strain MP_1, which is a lysing strain, the thin hyphal wall is disrupted and virus particles are liberated.

In the strain MP_2, hyphal cells were also invaded by a large number of particles localized in the cytoplasm. In some cells a release of fibrous material was observed, which formed a network from the core region of empty particles (Fig.3). Similar observations were made by Yamashita et al. (1973) in *Penicillium chrysogenum*.

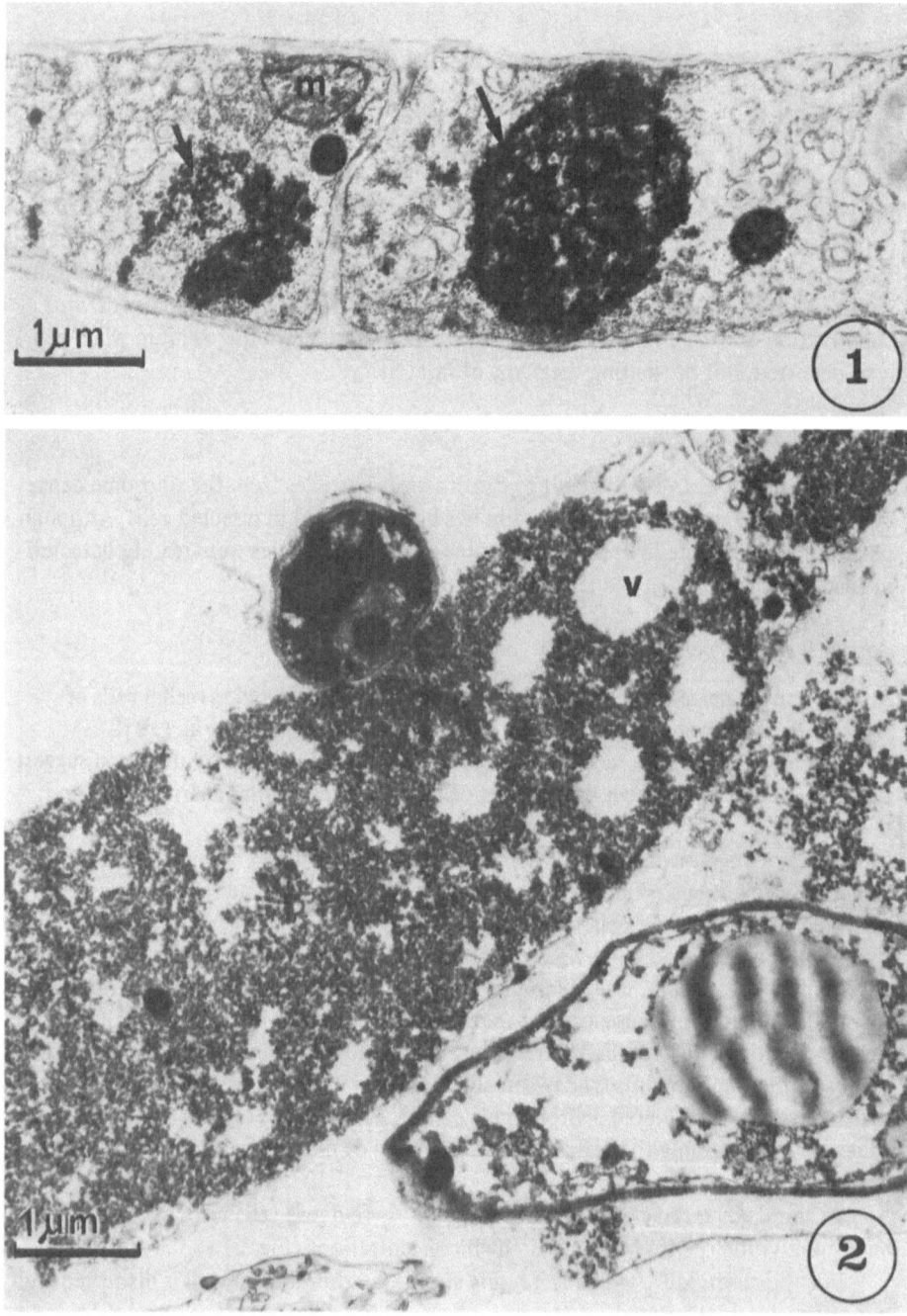

Fig. 1. *Mycogone perniciosa*. Rounded masses of granular material and virus particles (*arrow*)
Fig. 2. *Mycogone perniciosa*. Hypertrophied hypha of strain MP$_1$ filled with 42 nm virus particles
m = mitochondria; V = vacuole

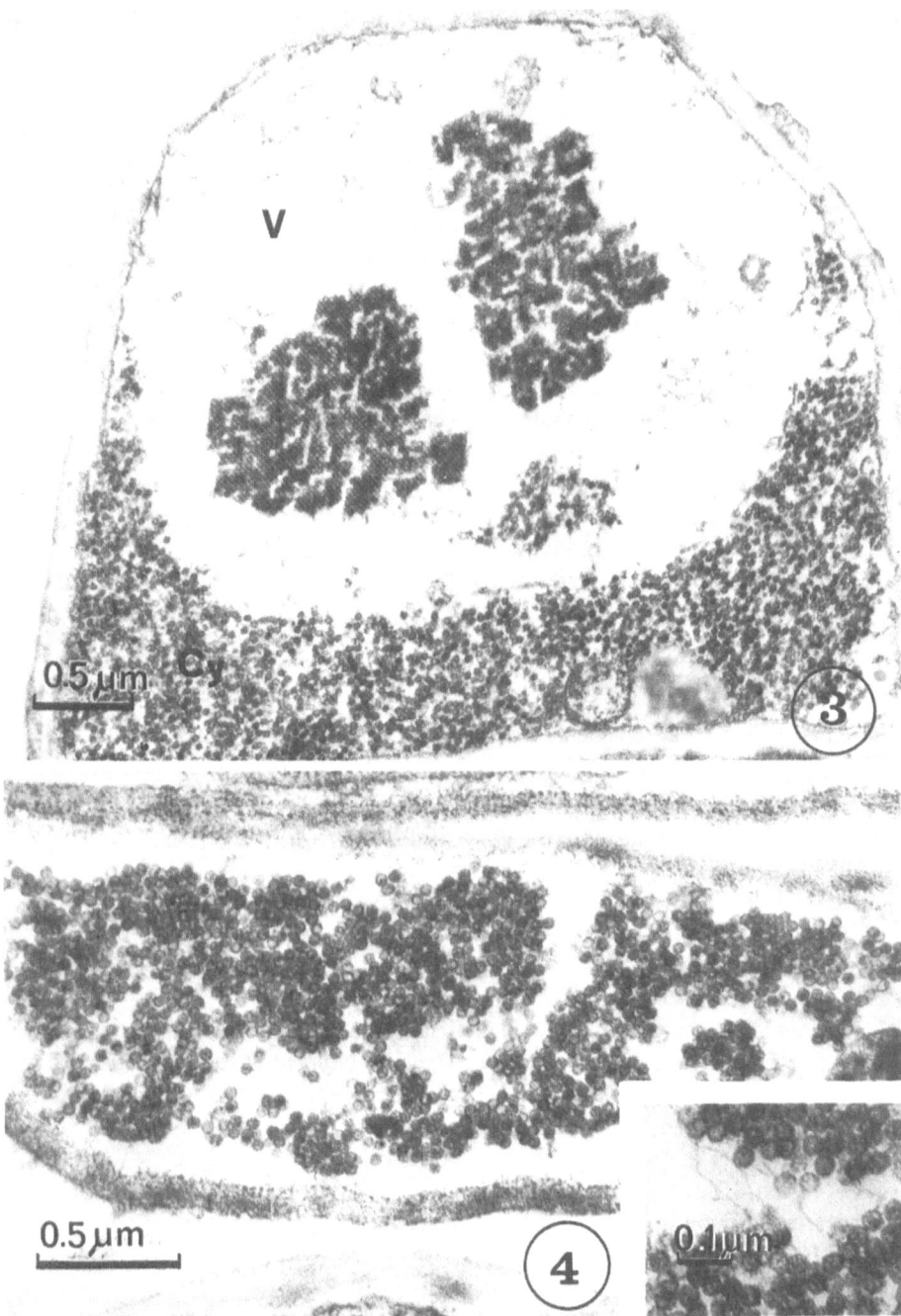

Fig. 3. *Mycogone perniciosa*-strain MP₂. Crystal arrays in large vacuole. Free particles in peripheral cytoplasm

Fig. 4. *Mycogone periniciosa*-strain MP₂. Free intracytoplasmic virus particles in a lysed cell. *Inset,* Fibrous material released from core region of empty protein shell

V = vacuole; *Cy* = cytoplasm

However, with the MP$_2$ strain there often occurred a spontaneous crystallization of virus particles in the central vacuoles. These crystals represented densely packed arrays of particles, with different complex patterns (Fig. 4).

Tubular formations of the same diameter as virus particles were also observed.

Cytological Aberrations. A characteristic of infection was the occurrence of numerous aggregates of dense lamellar inclusions in the cytoplasm of most infected cells.

They were dispersed throughout a large part of the cytoplasmic area. These components appeared in different configurations according to the angle of sectioning: as straight lines or lightly curved, rolled into scrolls, in rings, arcs or helicoidal structures (Figs. 5, 6).

Virus particles were usually interspersed between the inclusions. Some of the particles were lying parallel to the alternating layers, others occurred regularly side by side along the inner part of a filament.

It was also observed in one side of the inclusions, that regularly spaced leaves, like ribosomes but smaller, occurred.

None of these structures was observed in any of the noninfected tissues, and they are therefore considered as virus-associated components.

These virus-specific infection products were similar in several respects to the intracellular cylindrical inclusions induced by Potyviruses (Lawson et al., 1971) and it may be supposed that they are related directly to the virus multiplication and are not a byproduct of the infection.

Occurrence of Virus Particles in Aleurospores. Fungal spores of *Mycogone perniciosa* have two cells: a basal cell and a round terminal cell with a dark thick irregular wall:

The virus particles were generally found in very high concentrations in both cells. In the basal cells many crystals occurred. In the upper part vacuoles and cytoplasm were filled with numerous particles. Intracytoplasmic filamentous structures such as loops and circles were also seen.

Very rarely rows of rod-shaped viruses 18 x 120 nm were observed in vacuoles.

3.2 Virus-Infected Mushroom

Mushroom viruses have been detected in ultrathin sections of diseased vegetative mycelium, fruiting bodies and basidiospores of *Agaricus bisporus*. Several isometric (25 nm to 34 nm) and bacilliform virus particles have been identified (Dieleman-van Zaayen, 1972, 1975; Dieleman-van Zaayen and Igesz, 1969).

The presence of several isometric mushroom viruses (MV$_1$: 25 nm, MV$_2$: 29 nm, MV$_4$: 34 nm, MV$_5$: 50 nm) was previously reported (Albouy, 1972). In addition a new particle, 70 nm in diameter with a tubular caudal appendage was found (Albouy et al., 1973).

3.2.1 Intracellular Appearance of the Main Types of Known Mushroom Viruses – MV$_1$ – MV$_2$ – MV$_4$ – MV$_5$

The different isometric viruses could be distinguished from each other on the basis of their location, accumulation, and some aspects of their cellular effects.

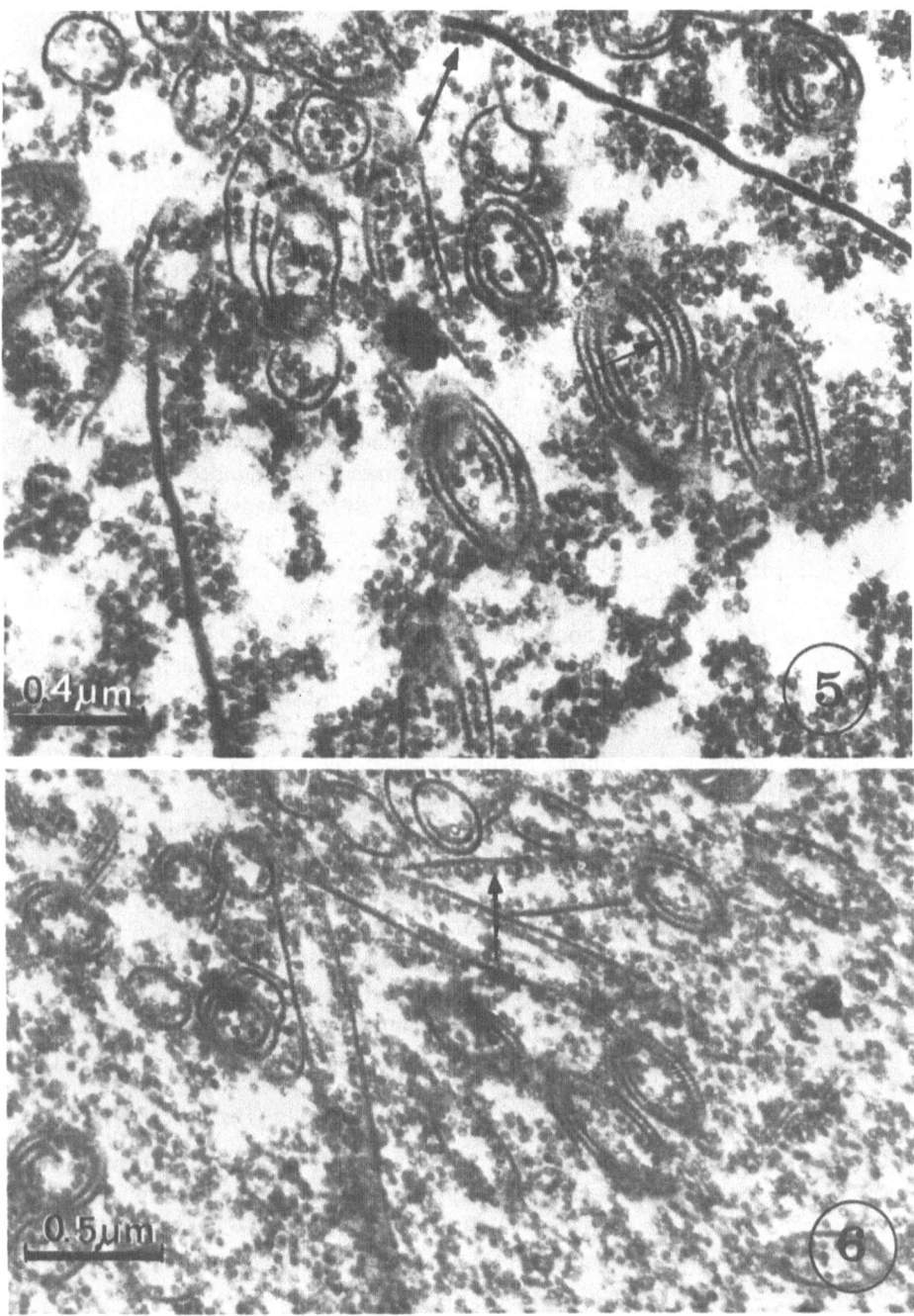

Figs. 5 and 6. *Mycogone perniciosa*. Different configurations of lamellar inclusions with a row of virus particles (*wide arrow*). Note the ornamentations along the inner part (*thin arrow*)

Viruses accumulated in different ways in the cytoplasm:
— in cytoplasm mixed with ribosomes
— in well defined areas in which the viruses were concentrated and apparently free
    from other constituents
— or compartmentalized within a membrane.

The most noticeable ultrastructural changes were a faster lysis of cells with a
marked decrease in cytoplasm and organelles, and a prominent development of cytoplasmic membranes. In young cells no marked changes were observed in nuclei or
mitochondria.

The four main types of particle were usually in combination, but MV_2 and MV_4
seemed the most numerous and were observed in every section of diseased mushrooms.
The MV_5 particles were only occasionally detected in ultrathin sections.

Aggregates of MV_1 contained few particles and were surrounded by a cytoplasmic
membrane. They were clustered in rings or loose aggregates always isolated from the
cytoplasm (Figs. 7, 8). We also found in cells infected by MV_1 single-membrane vesicles
containing electron dense bodies similar to ribosomes. These rounded vesicles occurred
in groups of 2 to 8, each group surrounded by a single membrane.

The proportion of cells infected by MV_2 was high, but each cell contained only
a few particles. They were scattered randomly throughout or clumped together in
small groups within the cytoplasm (Fig. 9). They were never enclosed in a membrane.

According to the observations of Dieleman-van Zaayen (1972, 1975) the occurrence of MV_4 was predominant. These particles can form dense round aggregates in
young cells associated with numerous developing membranes of the endoplasmic reticulum or are grouped near a dolipore (Fig. 11). They often appeared in small quantities,
enclosed in a loose single bounding membrane which appeared to isolate the aggregates
of empty particles from the rest of the cytoplasm (Fig. 10).

In older cells the large central vacuole contained virus particles which were closely
aggregated into a "stick", ring or cylindrical array. These formations were also reported
in vacuoles of *Penicillium chrysogenum* by Yamashita et al. (1973).

Few cells were infected by MV_5, but the intracytoplasmic inclusion of this virus
contained large numbers of particles (a few hundred per cell). They were often seen in
combination with MV_4 particles in a single vacuole (Fig. 11).

These MV_5 particles are known to be uncommon and to occur in low concentrations (Hollings et al., 1968; Hollings, 1978).

3.2.2 A New Morphological Viruslike Particle Associated with Pathogenic Effects

The presence of tailed viruslike particles of completely different morphological type
was observed in carpophores of *Agaricus bisporus* presenting severe abnormalities of
"teratologie" (Albouy and Lapierre, 1972). Similar particles associated with another
type of symptom were also described by Lesemann and Koenig (1977).

Strongly cytopathic effects — alterations of cytoplasm, nucleus, mitochondria, and
elaboration of filamentous structures — were studied.

Morphology of Particles. Complete structure of individual particles was not resolved
in detail but the morphology could be inferred from an examination of numerous
sections and particles. The rounded particle 70 nm in diameter contained a double

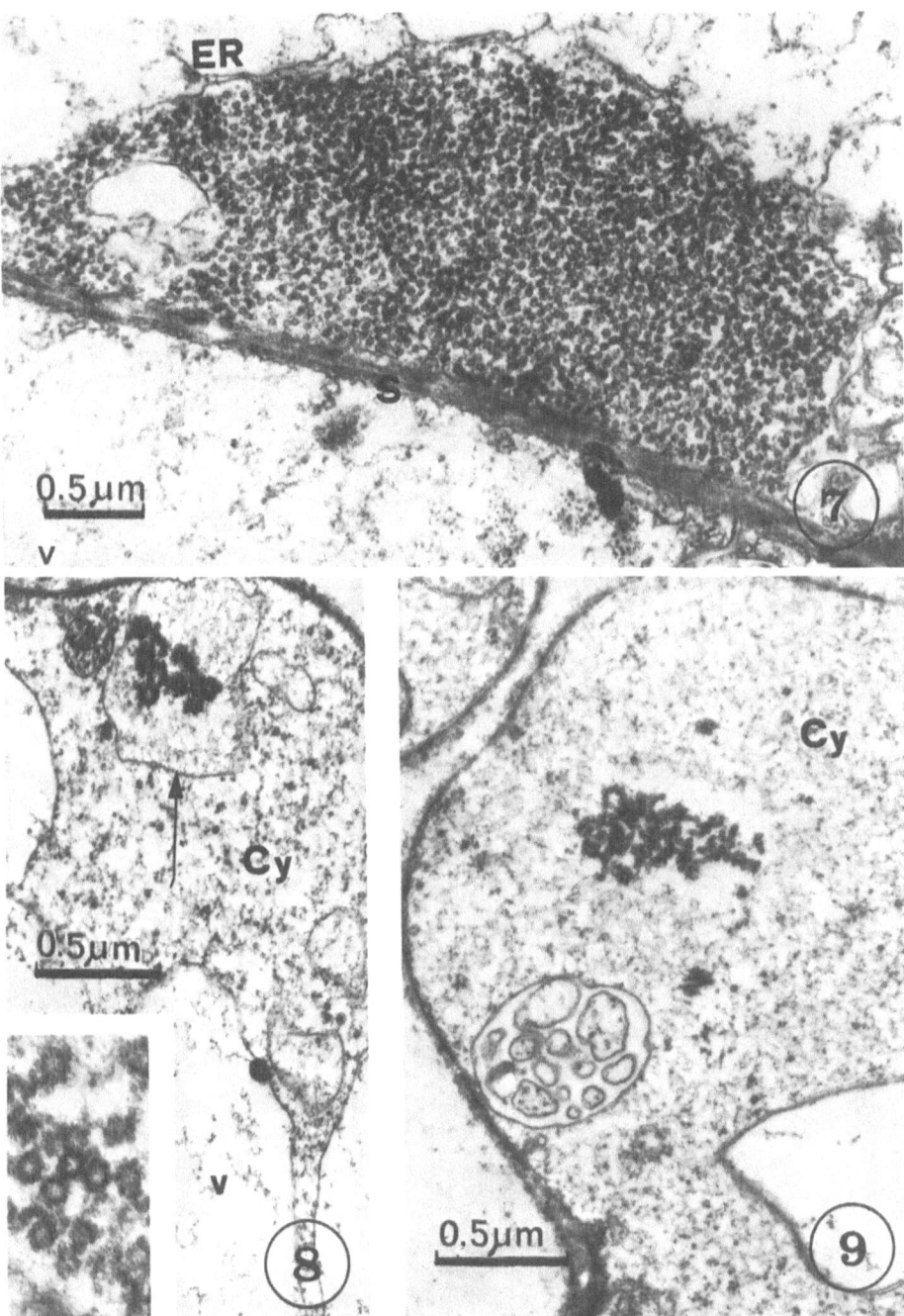

Fig. 7. Numerous MV_5 particles grouped along a septum

Fig. 8. Compartmentation of aggregated MV_1 particles. *Inset*, Ring aggregate

Fig. 9. MV_3 particles in a free cytoplasmic area

ER = endoplasmic reticulum, Cy = cytoplasm, V = vacuole

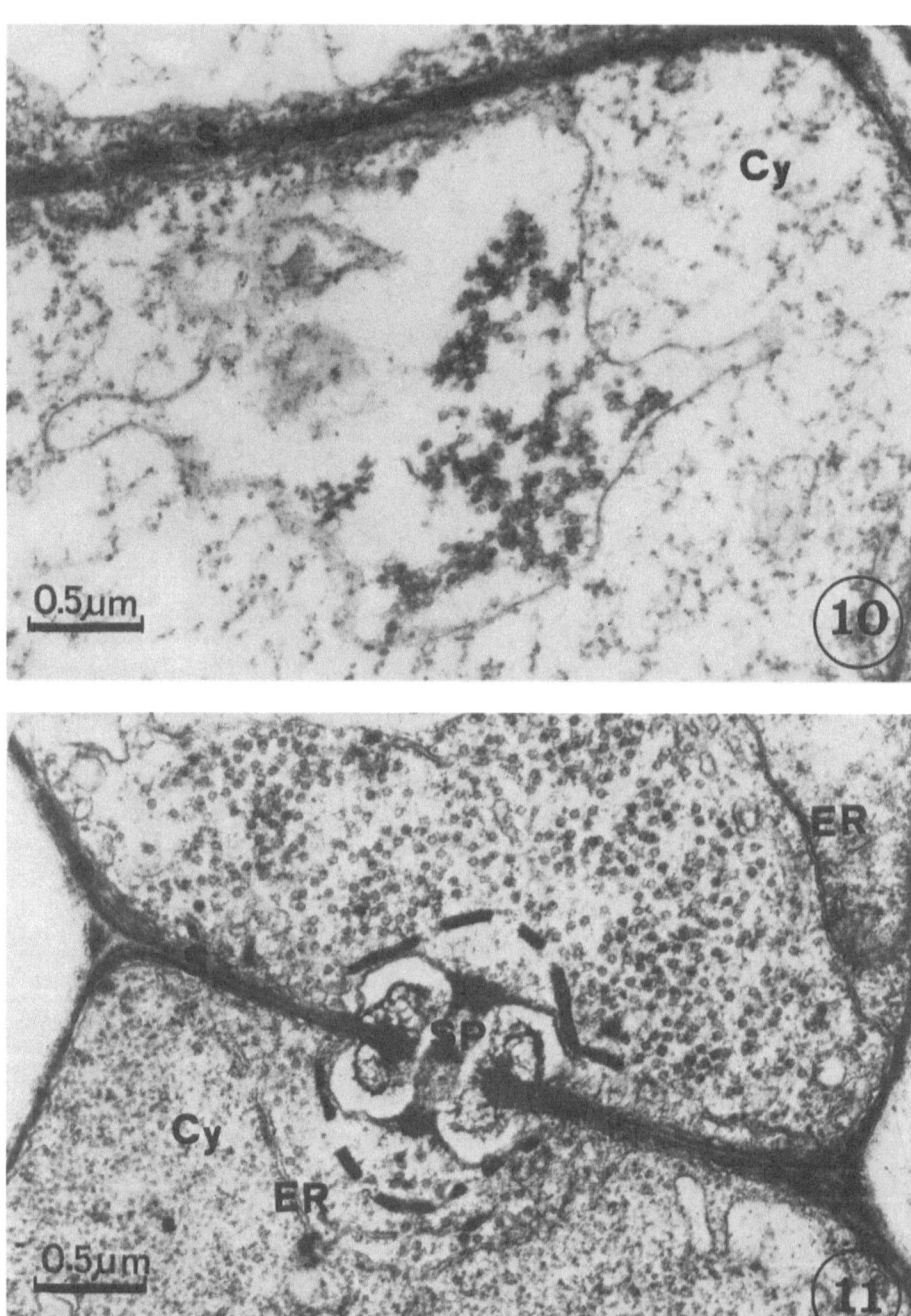

Fig. 10. Loose single membrane enclosing MV$_4$ particles (empty protein shell)
Fig. 11. Intracytoplasmic MV$_4$ particles near a dolipore (septal pore = *SP*)
Cy = cytoplasm, *ER* = endoplasmic reticulum

membrane envelope: two dark layers separated by a lightly stained layer. The outer layer appeared darker. The dense core varied in morphology. Most cores appeared to be homogeneous and electron-dense. A few were electron-lucent, and many were intermediate in density. The space inside the particle between the core and the envelope was more or less transparent and granular (Fig. 12).

Some of the particles appeared to have a "tail", like a caudal appendage variable in length (150–200 nm), thickness (25 nm) and form. This tail could be straight or a sinuous vesiculated tube (Fig. 13). The head appeared often slightly oval and the dark core seemed to be prolongated into the tail. Lesemann (1977) has reported such tailed particles ("club-shaped particles") in ultrathin sections and in negatively stained crude extracts or partially purified preparations of fruiting bodies.

It is possible that such tailed particles represent some morphological stage.

Alterations of Cytoplasm. The most notable difference between infected and uninfected tissues was the general increase in density of the cytoplasm, and the prominent cytoplasmic membranes. From these membranes numerous vacuolelike vesicles apparently coalesced to form larger vacuoles. They contained granular material and fibrils.

Within this strongly vesiculated cytoplasm a notable feature was the presence of a labyrinth of membranes within membranes forming smooth-surfaced tubules (Fig. 13). Dark material accumulated in these invaginations of cytoplasmic membranes. Particles of 70 nm are scattered in the vacuolized cytoplasm but they are never seen in the vacuoles. The cells appeared extremely disorganized and the vesiculation apparently represents a response to the infection by the tailed particles.

Alterations of Mitochondria. Several observations lead to the conclusion of the association of mitochondria with infection. This organelle exhibited some morphological changes. Aggregates of mitochondria (3 to 8) were often observed in certain areas of the cytoplasm of infected cells. These mitochondria always appeared with an unusual homogeneous and electron transparent material and seemed hypertrophied. The surface of the mitochondria was frequently covered with ribosomes (Fig. 14).

A striking feature was the deformation of the mitochondria. At one pole the double membrane was altered and the outer layer disappeared as the inner was elongated and convoluted. In this opened portion, rounded or tailed particles were observed (Figs. 15, 16).

The accumulation of particles in the mitochondrial space suggests that the mitochondria may play a role in virus synthesis.

Alterations of Nucleus. Another characteristic of infected cells is abnormal nuclei. The two layers of the nuclear membrane were separated at intervals and corrugated. The nucleolus was dark and prominent (Fig. 18) and several mitochondria were generally grouped around the nucleus.

Degradation of the nuclear unit membranes produced a "loose" envelope.

Occurrence of Filamentous Structures. Perhaps the most conspicuous and constant aspect of infection was the occurrence of aggregates of thin lamellar filamentous structures, which varied in their appearance depending on the plane of sectioning. They consisted

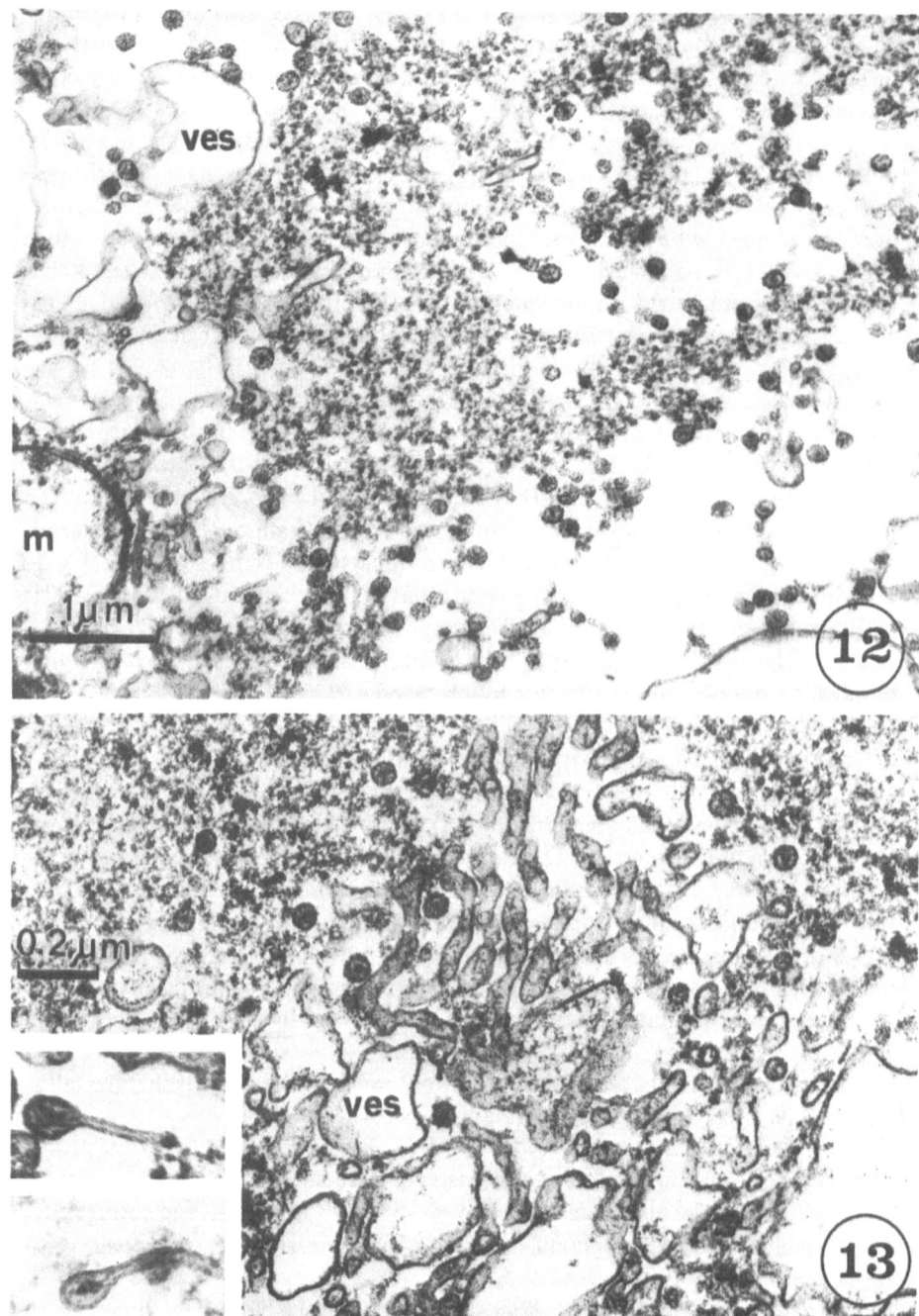

Fig. 12. Numerous 70 nm particles dispersed in the cytoplasm

Fig. 13. Vesiculated cytoplasm with smooth-surfaced tubule among rounded particles. *Inset,* Tailed viruslike particles

ves = vesicle, *m* = mitochondria

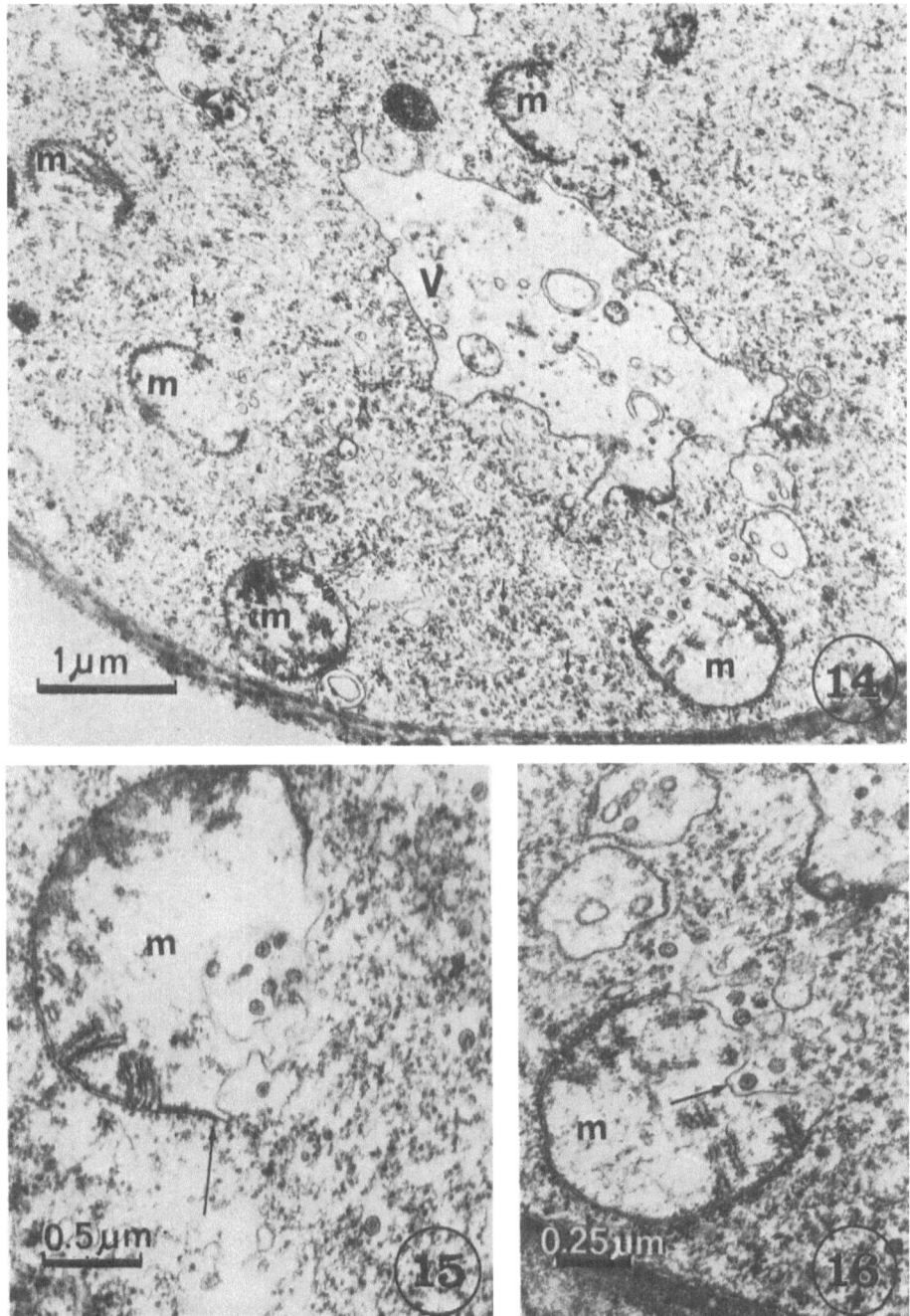

Fig. 14. Part of an infected cell with a dense cytoplasm and altered mitochondria
Fig. 15 and 16. Detail of altered mitochondria. Note close to the inner convulated membrane, rounded particles (*arrow*)
m = mitochondria, *V* = vacuole

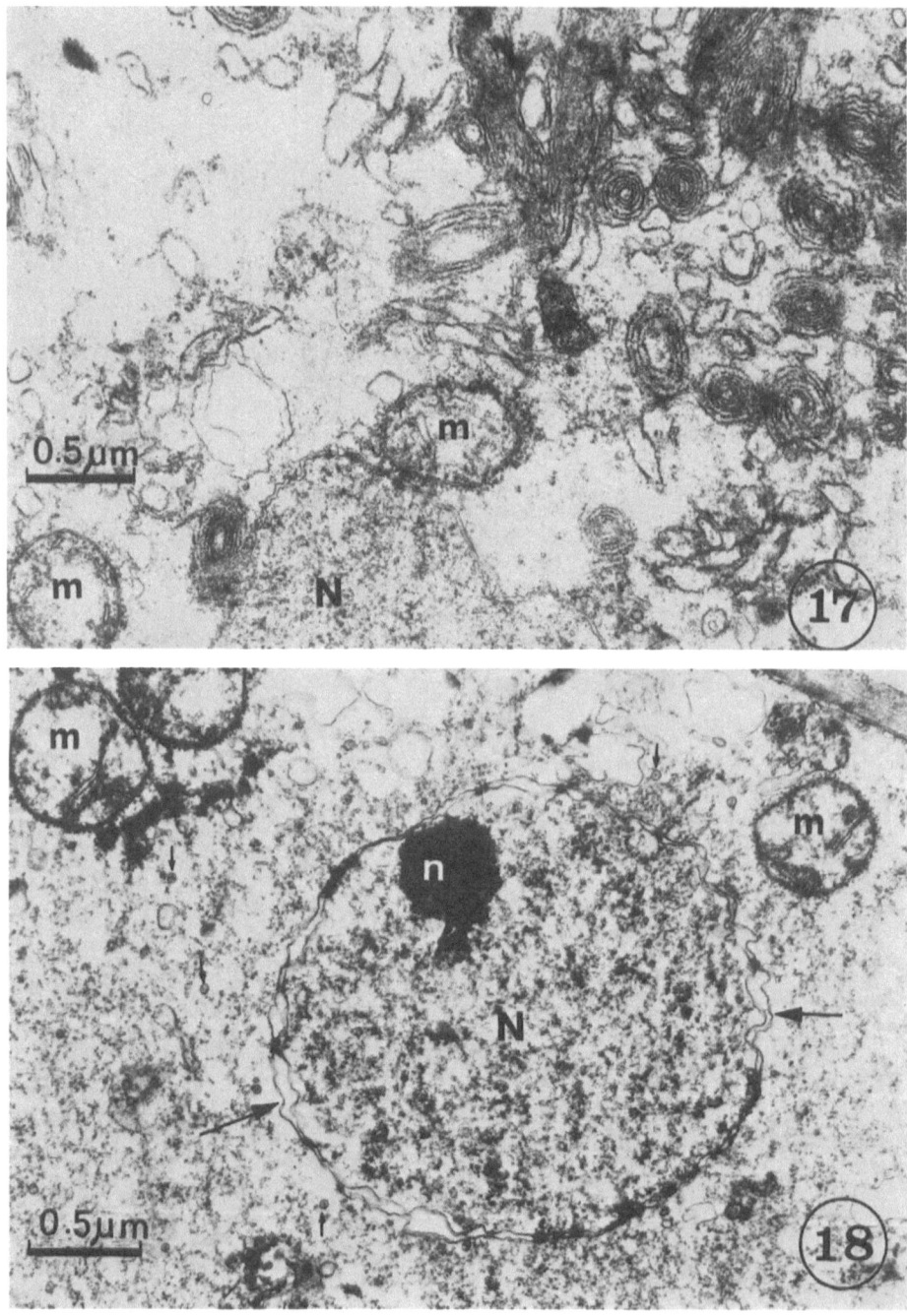

Fig. 17. Part of a cell with dense-packed filamentous structures
Fig. 18. Loose nuclear envelope of an altered nucleus
m = mitochondria, N = nucleus

of very dense fibrous lamellar material in regular arrays, often in concentric multi-layered structures (2 to 6 filaments), and appeared spindle and oval-shaped in longitudinal or cross sections (Fig. 19).

Usually they were clustered in certain areas of the cytoplasm, often in the center of the cell. They occurred as profuse stacks of lamellae and several were grouped in the same orientation, forming dense areas of numerous close-packed configurations (Fig. 17).

Filamentous structures were also seen in cells with normal dense cytoplasm and with scattered particles, as well as in cells where organelles and structures were disorganized (Fig. 20). The function of such filaments is unknown, they may represent a secondary cellular response to virus infection.

4 Conclusion

Interactions between viruses and susceptible cells result in cellular alterations (cytopathic effects). These morphological changes varied very much in severity with the host and the virus. Some host alterations are a general response to infection; like proliferation of membranes, concentration of ribosomes, accumulation of particles etc. . . . but others are specific and similar to responses to plant pathogenic viruses. For example vesicles in chloroplasts are typical alterations induced by the Tymoviruses (Lesemann, 1977), pinwheel inclusions are observed with Potyvirus infections and inclusion bodies are formed by Caulimoviruses (Lawson et al., 1971; Lawson and Hearon, 1977).

These cytopathological data (prominent membranous material, membrane-bound vesicles, general cellular disorganization) suggest that some modifications induced by fungal viruses are common in plant virus infections and may be significant in virus synthesis. Other modifications may be characteristic of a virus group or of individual viruses (lamellar and filamentous inclusions).

Virus cytopathic effects may lead to the death of cells by inhibition of host macromolecular synthesis, alteration of organelles (like mitochondria), or cell membranes (lysis) and accumulation of byproducts of infection. Such cellular effects by fungal viruses may be related to other damage such as reduced growth of mycelium, sporophore abnormalities (70 nm virus particle in *Agaricus bisporus*), and general crop losses.

Acknowledgments. We are greatly obliged to Dr. M. Favre Duchartre (Lab. Bot., Fac. Sci. Reims) who allowed us to carry out this study. We thank Mr. J.C. Poutier for technical assistance.

Summary

Electron microscopic studies of different fungal viruses (*Ophiobolus graminis*, *Sclerotium cepivorum*, *Mycogone perniciosa*) and of the mushroom viruses of *Agaricus bisporus* showed various aspects of their intracellular appearance, location and sequences of the ultrastructural changes caused by the virus infection.

In ultrathin sections of sporophore tissues of *Agaricus bisporus* we detected six types of virus particles. They are distributed as individuals or aggregates in the cytoplasm or vacuoles. Virus 2 had a fast cytophathic effect with degeneration or disappearance of cytoplasm and organelles. Some cylindrical arrays occurred with virus 1 and 4.

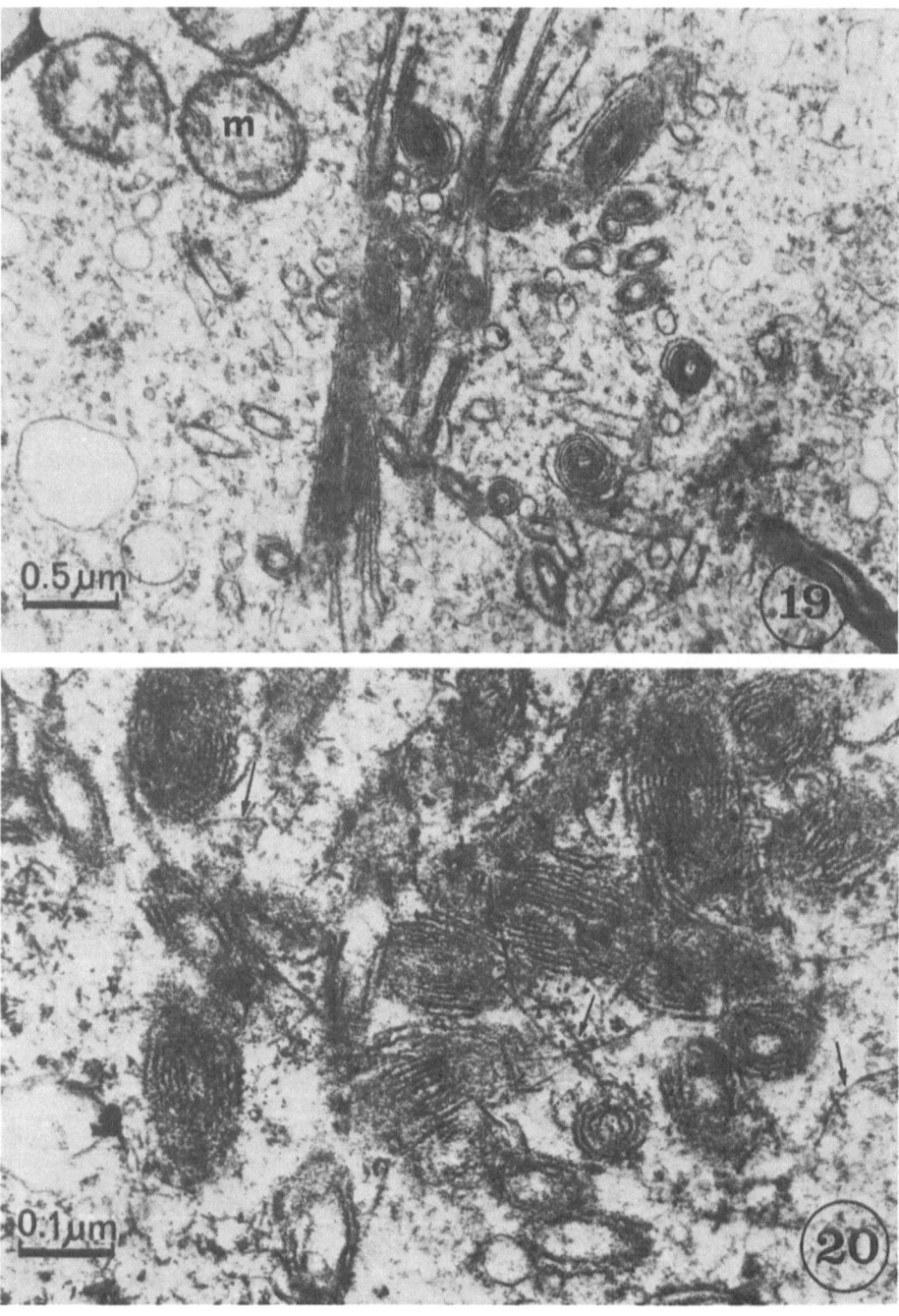

Fig. 19. Different configurations of filamentous structures
Fig. 20. Dissociated filaments in cross or longitudinal section (*arrow*)
m = mitochondria

A new morphological type of mushroom virus was also described: 70 nm particles with an electron-dense core and a membranelike envelope. Some of them showed a taillike appendage. The cytoplasm of cells contained dense lamellar inclusions and membraneous structures. Some figures suggested also association of particles with mitochondria.

In *Mycogone perniciosa*, we observed first an accumulation in the cytoplasm of dense granular material leading to spherical aggregates which fill the cell. Virions appeared embedded in this matrix, in great concentration. Another remarkable observation in sections of *Mycogone perniciosa* was the presence of numerous lamellar structures (in parallel arrays, rings, arcs . . .) accompanying the virus particles.

Two types of particles (30 and 45 nm) in *Sclerotium cepivorum* induced development of cytoplasmic membranes and vesicles followed by a complete disorganization of the cell.

References

Albouy J (1972) Etude ultramicroscopique du complexe viral de la "goutte sèche" de carpophores d'*Agaricus bisporus*. Ann Phytopathol 4: 39–44

Albouy J, Lapierre H (1971) Quelques aspects de l'infection virale chez les champignons (*Sclerotium cepivorum, Ophiobolus graminis, Agaricus bisporus*). Ann Univ A R E R S 9: 333–339

Albouy J, Lapierre H (1972) Observations en microscope électronique de souches virosées de *Mycogone perniciosa* (Magn) Cost et Dufour agent d'une môle du champignon de couche. Ann Phytopathol 4: 353–358

Albouy J, Lapierre H, Molin G (1973) Mise en évidence d'un nouveau type de particules dans des hyphes de carpophores d'*Agaricus bisporus*. C R Acad Sci Paris Sér D 276: 2805–2807

Dieleman-van Zaayen A (1972) Intracellular appearance of mushroom virus in fruiting bodies and basidiospores of *Agaricus bisporus*. Virology 47: 94–104

Dieleman-van Zaayen A (1975) Electron microscopy of virus-infected cultivated mushroom. Rep Tottori Mycol Inst 12: 139–150

Dieleman-van Zaayen A, Igesz O (1969) Intracellular appearance of mushroom virus. Virology 39: 147–152

Hollings M (1978) Mycoviruses: Viruses that infect fungi. Adv Virus Res 22: 1–51

Hollings M, Stone OM, Atkey PT (1968) Mushroom viruses. Rep Glasshouse Crops Res Inst 1967, p 101

Lapierre H, Albouy J, Faivre-Amiot A, Molin G (1971) Mise en évidence de particules virales dans divers champignons du genre *Sclerotium*. C R Acad Sci Paris Sér D 272: 2848–2851

Lapierre H, Faivre-Amiot A, Kusiak C, Molin G (1972) Particules de type viral associées au *Mycogone perniciosa* Magnus, agent d'une des môles du champignon de couche. C R Acad Sci Paris Sér D 274: 1867–1870

Lawson RH, Hearon SS (1977) Ultrastructure of extracted carnation etched ring virus inclusion bodies treated with proteolytic enzyme and DNAase. Phytopathology 67: 1217–1226

Lawson RH, Hearon SS, Smith FF (1971) Development of pinwheel inclusions associated with sweet potato russet crack virus. Virology 46: 453–463

Lesemann DE (1977) Virus group-specific and virus-specific cytological alterations induced by members of the Tymovirus group. Phytopathol Z 90: 315–336

Lesemann DE, Koenig R (1977) Association of clubshaped virus-like particles with a severe disease of *Agaricus bisporus*. Phytopathol Z 89: 161–169

Yamashita S, Doi Y, Yora K (1973) Intracellular appearance of *Penicillium chrysogenum* virus. Virology 55: 445–452

Interactions of Fungal Viruses and Secondary Metabolites

R.W. DETROY and K.A. WORDEN

Northern Regional Research Center, Federal Research, Science and Education Administration, U.S. Department of Agriculture, Peoria, IL 61604/USA

Mycoviruses have been demonstrated in all major filamentous fungi groups and biochemically characterized from a number of fungi (Bozarth, 1972; Lemke and Nash, 1974). In view of the agricultural and industrial impact of a number of these virus-containing fungi, studies have been pursued during the past few years on how these viruses, in some fashion, may regulate metabolic products, such as antibiotics and other secondary metabolites. Current information will be reviewed involving the biological implications of viruses in fungi, especially their interactions with fungal metabolites in vegetatively growing tissue. Recent findings will be discussed relative to the physiological roles of viruses and their interaction with fungal metabolites.

1 Fungal Viruses and Host Metabolism

Shortly after the discovery of viruslike particles (VLP) in fungi and their subsequent characterization in certain *Penicillium* species (Ellis and Kleinschmidt, 1967; Banks et al., 1968), considerable attention was focused upon their role in fungal metabolism, especially antibiotic and mycotoxin synthesis. Fungi are capable of the synthesis of a wide variety of ubiquitous secondary metabolites (Birkinshaw, 1965) of which little is known regarding their genetic regulation. Lemke and Ness (1970) suggested that such compounds may be related to the presence of virus in fungal tissue. Evidence compiled to date indicates that fungal viruses are not directly involved in the synthesis of secondary metabolites. Current information on the implications of virus interactions with secondary metabolism is depicted (Table 1).

With regard to penicillin-producing fungi, all strains of *Penicillium chrysogenum*, regardless of virus titer, retain the ability to produce antibiotic (Lemke et al., 1973). Other antibiotic-producing fungi, *P. notatum* and *Cephalosporium chrysogenum*, contain viruses (Hollings and Stone, 1971; Lemke and Nash, 1974). Virus particles have also been demonstrated in a non-aflatoxin-producing strain of *Aspergillus flavus;* however, a more extensive investigation has not revealed virus in other aflatoxin-producing and negative strains of *A. flavus* (MacKenzie and Adler, 1972). Wood and Bozarth (1973) in subsequent work with the virus from the infected *A. flavus* strain showed that the particles contained no detectable nucleic acid.

In other virus-metabolite interactions, Detroy et al. (1973) describe secondary metabolite production associated with the absence of viruses. Strains of *Penicillium stoloniferum* and *P. brevicompactum* that produce the antiviral metabolite mycophenolic

Table 1. Relationship studies: mycoviruses and host metabolism

Organism	Activity or metabolite	Reference
Penicillium chrysogenum	Penicillin [a]	Banks et al., 1968 Lemke et al., 1973 Normansell and Holt, 1978
Penicillium notatum	Penicillin	Volkoff et al., 1972 Hollings and Stone, 1971
Aspergillus flavus	Aflatoxin	MacKenzie and Adler, 1972 Wood and Bozarth, 1973
Penicillium stoloniferum	Mycophenolic acid	Detroy et al., 1973
Penicillium brevicompactum	Mycophenolic acid	Detroy and Still, 1975
Penicillium citrinum	Mycophenolic acid	Borré et al., 1971
Penicillium variabile		
Penicillium citrinum	Sporulation	Benigni et al., 1978

a   The mention of firm names or trade products does not imply that they are endorsed or recom-
mended by the U.S. Department of Agriculture over other firms or similar products not mentioned

acid do not contain viruses, whereas virus-containing strains do not produce the metabolite. Unrelated experiments by Borre et al. (1971) have shown that mycophenolic
acid inhibits the formation of lytic plaques by virus-infected cultures of two other
Penicillium species, *P. citrinum* and *P. variabile*. Benigni et al. (1978) recently characterized the fungal viruses (Pcit-1 and Pcit-2) and their dsRNA from *P. citrinum*. More
recently these authors have shown that the sporogenic parental isolate contains a mixture of two virus particles (Pcit-1 and Pcit-2), while the asporogenic strain shows almost
exclusively Pcit-1 particles. The interrelationships of mycophenolic acid and other fungal metabolites with regard to fungal virus replication and their conferrence of protection will be discussed later in detail.

2 Fungal Viruses and Plant Diseases

Viruses have been isolated in over 30 species of plant pathogenic fungi (Lemke, 1977)
of which a number have yielded evidence for viral dsRNA. Initial interest in these virus-
containing fungi evolved around the possibility of biological control of plant pathogens (Table 2). The pathogenicity of *Gaeumannomyces graminis* (Rawlinson et al.,
1973; Rawlinson and MacLean, 1973; Rawlinson et al., 1977), causative agent of
wheat take-all disease and *Periconia circinata* (Dunkle, 1974a,b), cause of sorghum rot
fungus, does not appear to be influenced by the presence of viruses. Virus particles were
isolated from pathogenic and nonpathogenic strains of *P. circinata;* hence, this virus is
not related to toxin production and pathogenesis by this fungus (Dunkle, 1974b).

    In studies involving the take-all fungus of wheat, *G. (Ophiobolus) graminis*, the
absence of correlation is less certain, particularly among strains studied in France

Table 2. Relationship studies: mycoviruses and phytopathogenic fungi

Organism	Activity	References
Gaeumannomyces (Ophiobolus) graminis (wheat disease)	Take-all disease	Lemaire et al., 1971 Lapierre et al., 1971 Rawlinson et al., 1973, 1977
Helminthosporium maydis (southern corn blight)	Toxin production	Bozarth, 1972
Periconia circinata (sorghum rot)	Toxin production	Dunkle, 1974a

(Lapierre et al., 1971; Lemaire et al., 1971). British investigators (Rawlinson et al., 1973; Rawlinson and MacLean, 1973; Rawlinson et al., 1977), however, have studied several varieties and reported inconsistency in their correlations between viruses and hypopathogenicity. These authors concluded that the two-sized viruses in their strains were relatively latent.

Initially, virus particles were found only in phytopathogenic strains of *Helminthosporium maydis,* the southern corn blight fungus; however, additional studies revealed the presence of viruses in both mildly and nonpathogenic strains of this fungus (Bozarth, 1972).

The remaining information discussed in this section represents recent findings on virus-containing fungal pathogens (Table 3).

Table 3. Current studies with virus-containing phytopathogenic fungi

Organism	Activity	References
Synchytrium endobioticum (potato wart)	Organelle disruption	Lange and Olson, 1978
Helminthosporium sp. (corn blight) India	Morphological plaques	Misra et al., 1978
Phytophthora sp. (potato blight)	Morphological VLPs	Styer and Corbett, 1978
Endothia parasitica (chestnut blight)	Hypovirulence	Day et al., 1977
Thielaviopsis basicola (root pathogen)	VLP	Bozarth and Goenaga, 1977

A systematic screening has been conducted in the genus *Fusarium* for viruses (Lapierre and Spire, 1974). Forty-five strains of *Fusarium oxysporum* were examined with only two strains containing VLP. Tests with *F. culmorum, F. avenaceum,* and *F. graminearum* yielded VLP only in the species *F. culmorum* (6 of 19 tested). Preliminary experiments with *Fusarium* species indicate that VLP have no effect on the pathogenicity of these fungi. Lange and Olson (1978) have recently described in ultrastructural studies VLP in different stages of the life cycle of the obligate potato pathogen,

Synchytrium endobioticum. A close association between VLP and disrupted organelles (mitochondria and lipid bodies) was observed. No VLP were observed in surrounding potato plant host cells.

Misra et al. (1978) have described morphological plaques in cultures of *Helmintho-sporium* spp. These plaques were serially inherited by new cultures and produced similar effects. Styer and Corbett (1978) have described two morphologically distinct types of VLP, one tubular and the other spherical, in nuclei of *Phytophthora infestans.* These viruses are exclusively intranuclear and occurred in the interphase and mitotic nuclei of hyphae, sporangia, and zoospores. Further studies with 25 new isolates showed that all possessed intranuclear VLP, some contained tubular VLP and others spherical viruses. The presence of tubular or spherical VLP was not correlated with growth rates, pathogenicity or other physiological parameters.

Day and coworkers (1977) have shown that all confirmed hypovirulent strains of *Endothia parasitica* (chestnut blight) carry dsRNA. No dsRNA was found in any of 15 pathogenic wild-type strains. The dsRNA is transmitted by hyphal anastomosis. No conclusions were drawn about the relationship between hypovirulence and the presence of dsRNA.

3 Interaction of Secondary Metabolites and Fungal Virus Replication

In limited studies that have been conducted to date, little evidence exists that fungal viruses play any direct role in regulation or production of fungal metabolites. However, more salient points of interaction between host metabolism and mycoviruses concern the role fungal metabolites (exclusive of killer toxins) may play in (1) the control of virus replication and, therefore, the preservation of the virus in certain fungal species; and (2) the natural protection for fungi against virus infection and proliferation.
A wide variety of secondary metabolites are produced by fungi possessing a myriad of biological activities against eukaryotic and prokaryotic organisms, protozoa, and animal and plant viruses.

Experiments in our laboratory on fungal viruses over the past few years center around the regulatory role that various biologically active, naturally occurring fungal metabolites play in the proliferation of fungal viruses in actively metabolizing fungal tissue.

4 Materials and Methods

The virus-containing strain of *P. stoloniferum* NRRL 5267 (ATCC 14586) and virus (-) strain NRRL 859 were supplied by the Agricultural Research Culture Collection (NRRL) maintained at the Northern Regional Research Center. During our investigations, the organism was maintained on M40Y (malt, yeast extract, sucrose and agar) slants.

Conidia ($10^6 - 10^7$) in 0.01% Triton X were used to inoculate each 500-ml Erlenmeyer flask containing 100 ml of yeast extract sucrose (YES) (2% yeast extract and 10% sucrose) medium. Incubation was at 26°C on a Brunswick rotary shaker at

250 r.p.m. for 48 h. For measurement of virus replication, 48-h mycelia were aseptically filtered and washed twice with sterile H_2O.

The washed mycelia were resuspended in sterile H_2O (5–7 g wet weight/100 ml) and 20-ml aliquots were transferred to sterile 300-ml flasks containing 50 ml H_2O plus additional carbon and/or nitrogen sources as indicated. Various isotopes were added and zero-time samples were removed for analysis.

Flasks were removed after 24 h at 26°C, pH taken and mycelia (1–1.5 g wet weight) were disrupted in a Bronwill mechanical cell homogenizer according to protocol described in Figure 1.

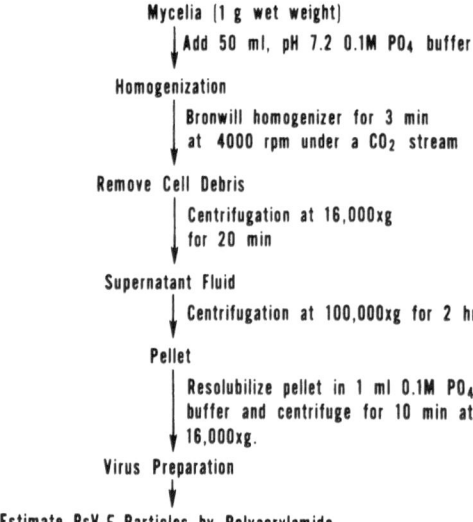

Fig 1. Protocol for isolating and estimating PsV-F particles in mycelia of *Penicillium stoloniferum* NRRL 5267

An aliquot of the cell-free supernatant fluid was treated with an equal volume of cold 5% TCA (in 0.1 M pyrophosphate) for 30 min which represented a measure of macromolecular synthesis. The precipitates were filtered on glass fiber discs, washed with cold 5% TCA, dried and placed in scintillation vials with 8 ml Aquasol cocktail for determination of radioactivity.

For estimation of in vivo virus replication, the remaining supernatant fluid was subjected to ultracentrifugation at 105,000 g for 2 h. The virus pellet was resuspended in PO_4 buffer and clarified by low-speed centrifugation. Polyacrylamide gels were prepared essentially as described by Loening (1967). Electrophoresis was carried out for 2.5 h at 7.0 mA/tube at 25°C. Gels were scanned at 260 nm by a Gilford linear transport system.

A PsV-F standard with 10.0 A_{260} units (absorbancy units at 260 nm in a 1-cm cell) per ml was prepared by sucrose gradient centrifugation and analyzed by gel electrophoresis for purity. A PsV-F standard curve was determined since the area under the PsV-F peak showed a linear relationship to amounts of PsV-F (0.03 and 0.25 A_{260} units) applied to the gels. Quantities of virus in mycelial homogenates were estimated by comparison of the area under the PsV-F peaks to known A_{260} units of standard PsV-F. Figure 2 depicts a typical profile of the fast- and slow-moving viruses obtained from *P. stoloniferum* tissue.

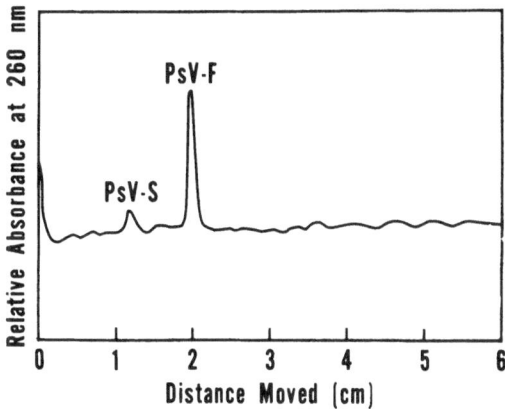

Fig. 2. Electrophoretogram of PsV-S and PsV-F from *Penicillium stoloniferum*. The 2.4% gels were run at 6.5 mA/tube for 2.5 h

For determination of viral dsRNA synthesis, gels were sliced and incubated in 2% periodic acid for 30 min and the radioactivity in the virus peaks determined in a Packard Scintillation counter. Viral RNA synthesis was measured as a function of ^{3}H uridine incorporation into the PsV-F peaks by gel electrophoresis.

5 Results and Discussion

A typical in vivo replication profile for viral dsRNA (both PsV-F and PsV-S) and fungal growth is depicted in Figure 3. Conidia germination in the YES medium begins at 16 h, followed by a rapid burst in macromolecular synthesis. The viral dsRNA was initially detected at 30 h, followed by a steady increase in dsRNA levels through 155 h incubation. The viral dsRNA increased with a concomitant increase in host protein, DNA and RNA synthesis and continued well after primary host macromolecular synthesis had reached maximal levels between 60 and 100 h growth. Similar experiments with a virus-free *P. stoloniferum* strain (859) indicated that the growth rates and macromolecular synthesis patterns were similar to those of the virus-containing 5267 strain (Still et al., 1975).

An optimum in vivo replication period for PsV-F was selected to evaluate the effect of fungal metabolites on virus replication. The test system involves measuring PsV-F replication in vegetative mycelia between 48- and 96-h development (Detroy and Still, 1975). A typical profile of PsV-F replication and fungal growth is depicted in Figure 4. The rates of RNA and protein synthesis were linear throughout the 48- to 96-h period; dry weight biomass paralleled PsV-F replication.

Since we had reported earlier (Detroy et al., 1973) that mycophenolic acid (MA) affected mycovirus replication and that an inverse correlation exists between the

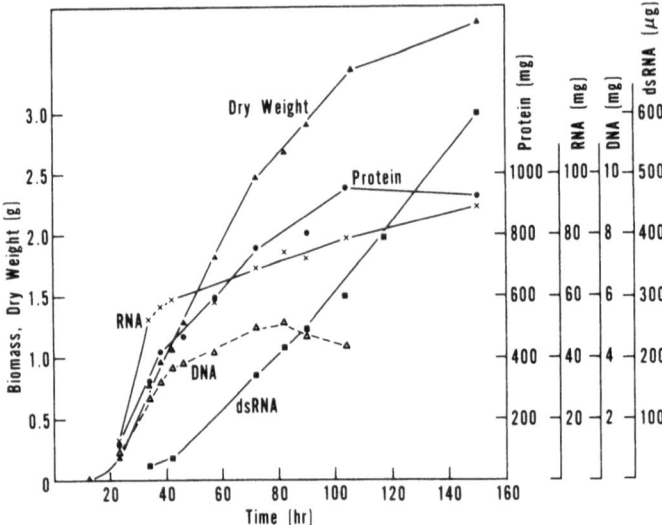

Fig. 3. In vivo viral dsRNA replication and host DNA, RNA and protein synthesis

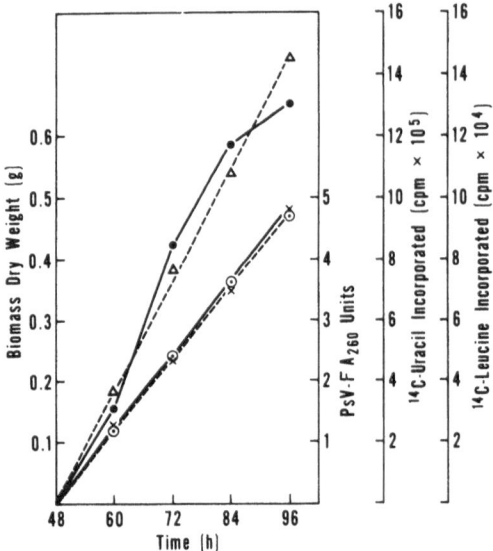

Fig. 4. Macromolecular and PsV-F synthesis in *Penicillium stoloniferum* vegetative mycelia. *Curves* show biomass increase, virus synthesis and uptake of ^{14}C precursors into acid-insoluble constituents in *Penicillium stoloniferum* cells. (x) Biomass, dry weight; (○) PsV-F synthesis; (△) ^{14}C-uracil incorporation; and (●) ^{14}C-leucine incorporation. Mycelia were analyzed at various intervals in 48- to 96-h incubation period

presence of virus and the production of MA (Table 4) in *P. stoloniferum* and *P. brevicompactum*, further studies with other fungal metabolites included MA.

Figure 5 depicts the structural-type metabolites that have proven effective as inhibitors of PsV-F in the in vivo test system. Mycophenolic acid was not effective against virus replication at levels below 150 μg/ml. MA at levels of 150 to 300 μg/ml resulted in a 31%–50% inhibition of PsV-F replication with minimal effects upon host protein and RNA synthesis.

Table 4. Biosynthesis of viruslike particles (VLP) and mycophenolic acid (MA) by strains of *Penicillium stoloniferum* and *Penicillium brevicompactum*

NRRL No.	MA, mg/g [a]	VLP [b]
P. stoloniferum		
5267	–	600
5275	12.0	ND [c]
5276	10.0	ND
859	11.0	ND
860	4.2	ND
861	4.5	ND
P. brevicompactum		
862	7.0	ND
2302	10.0	ND
2303	8.0	ND
5260	11.0	5.0
A-19207	12.0	ND
A-19377	11.0	ND
A-19378	18.0	ND
A-19379	14.0	ND
A-19380	8.1	ND
A-19383	7.0	ND
A-19462	6.5	ND
A-19464	14.5	ND
A-19466	15.5	ND
A-19467	16.3	ND

a   The mycelial dry weights are averages from duplicate 500-ml growth flasks inoculated with 10^6 spores/ml and harvested at 76 h
b   Analysis for VLP on extraction of 10 to 15 g dry weight mycelia
c   ND = not detected by electron microscopy or gel electrophoresis

Patulin, a fungal toxin produced by several species of *Aspergillus* and *Penicillium* (Hooper et al., 1944; Abraham and Florey, 1949), contains the α,β-unsaturated lactone structure. This toxin has antibiotic activity, is toxic to animals and is carcinogenic in rats and mice (Dickens and Jones, 1961, 1965; Norstadt and McCalla, 1969). As for mode of action, Jones and Young (1968) suggested that the SH group of thiol enzymes reacts with the double bond conjugated to the lactone carbonyl group of patulin, and Atkinson and Stanley (1943) suggested that the antibiotic activity of patulin may be due to its reaction with SH groups of microbial enzymes. Ashoor and Chu (1973) showed that patulin inhibits yeast alcohol dehydrogenase (ADH) and rabbit-muscle lactic dehydrogenase activities in vitro.

Preliminary results indicated that the mycotoxin patulin can block mycovirus replication. Patulin at 11, 16, and 20 µg/mg dry weight mycelia blocked PsV-F replication

Metabolite[a]	Concentration μg/ml medium	Inhibition (%)
Mycophenolic acid 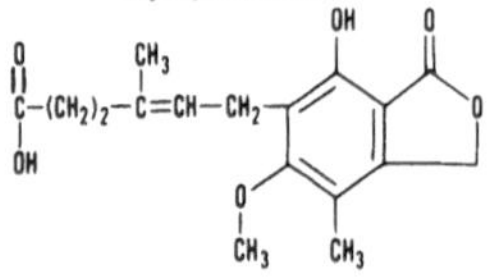	150-300	31-100
Patulin 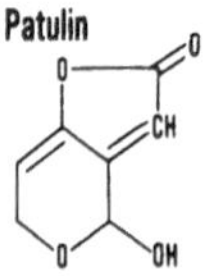	100	85
Gliotoxin 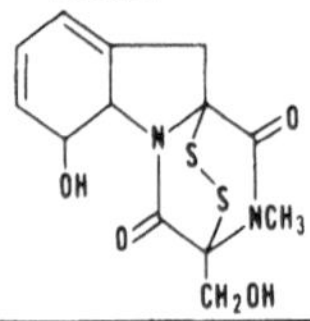	100	50

[a]Metabolites added to vegetative mycelia under proliferating conditions.

Fig. 5. Fungal metabolites inhibitory to PsV-F replication

26, 61 and 71%, respectively, compared with untreated controls. At these levels, host biomass RNA and protein synthesis were minimally affected. Nonproliferating fungal mycelium is capable of continued support of PsV-F replication, which is sensitive to patulin.

Gliotoxin, a known antiviral fungal metabolite, was most effective against PsV-F replication, 50%, at 100 μg/ml. Figure 6 depicts a number of other biologically active

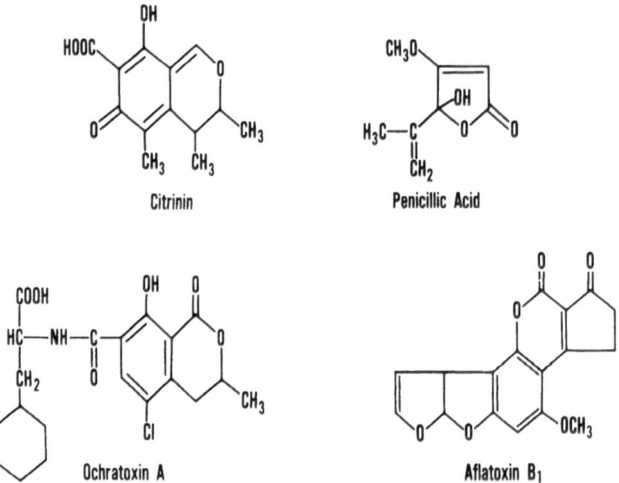

Fig. 6. Fungal metabolites: No inhibition to PsV-F replication

mycotoxins such as citrinin, penicillic acid, ochratoxin A, and aflatoxin B which exhibited no activity against the virus in vivo. Lack of biological activity of these metabolites may also result from the inability of fungal cells to take up these metabolites actively and from solubility problems.

Some variation in the PsV-F inhibitory levels and growth effects have resulted with the MA experiments at levels above 200 μg/ml with proliferating mycelia. Attempts to verify the MA effects in proliferating mycelia have been most erratic, partly due to solubility of the metabolite.

To avoid these difficulties, experiments have been conducted with washed mycelia incubated under nonproliferating conditions. In order to discern the more specific effects of MA, both PsV-F and viral dsRNA synthesis are measured in both proliferating and nonproliferating mycelia. Figure 7 indicates a comparison of ^{3}H-uridine incorporation into PsV-F viral dsRNA and PsV-F synthesis in proliferating and nonprolifer-

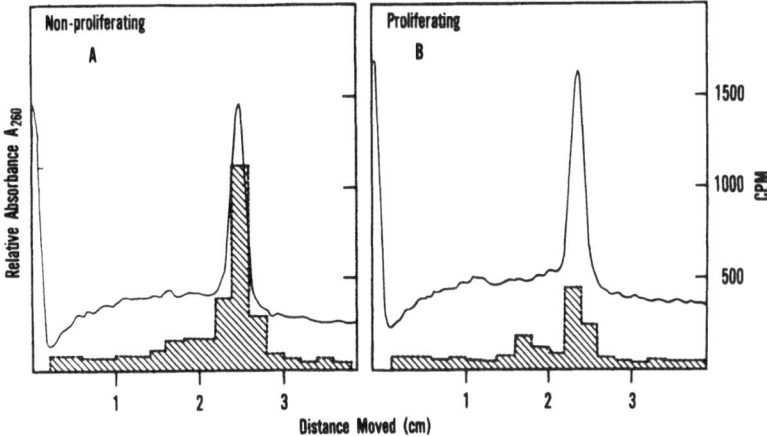

Fig. 7 A,B. PsV-F synthesis and ^{3}H-uridine incorporation into PsV-F dsRNA in proliferating and nonproliferating mycelia. A Nonproliferating: (-) A_{260} PsV-F units on polyacrylamide gels; (*shaded area*) ^{3}H$_{cpm}$ in PsV-F peak. B Proliferating: (-) A $_{260}$ PsV-F units on polyacrylamide gels; (*shaded area*) ^{3}H$_{cpm}$ in PsV-F peak

ating tissue. Optimal ^{3}H-uridine incorporation was obtained in the nonproliferating mycelia after 24 h compared to the proliferating mycelia. Figure 8 indicates the time course of PsV-F synthesis and ^{3}H-uridine incorporation into viral dsRNA. Biomass remains relatively constant throughout 48 h with little autolysis. PsV-F and viral dsRNA synthesis was maximal by 24 h and declined thereafter. For subsequent experiments, a 24-h nonproliferating assay was used to measure the effects of exogenously added mycophenolic acid.

Table 5 depicts the effect of MA upon PsV-F synthesis and ^{3}H-uridine incorporation into PsV-F dsRNA. At 100 μg/ml, MA had little effect upon PsV-F or dsRNA synthesis; however, levels of MA at 150 and 200 μg/ml resulted in a 33% and 74% PsV-F inhibition, respectively. PsV-F dsRNA at 200 μg/ml MA was dramatically reduced as

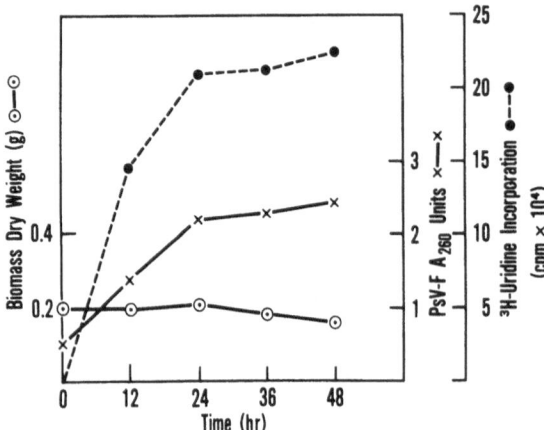

Fig. 8. Time course of PsV-F and ^{3}H-dsRNA synthesis in nonproliferating mycelia. ⊙——⊙ Biomass, dry weight (g); ●——● ^{3}H-uridine (cpm) in PsV-F peak; x——x A_{260} units of PsV-F

Table 5. Inhibition of PsV-F synthesis and dsRNA replication in nonproliferating mycelia of *Penicillium stoloniferum*

Addition	Biomass D.W. (g)	A_{260}	Percent inhibition	[3]H incorporation into PsV-F	
				cpm x 10[4]	Percent inhibition
Control, ot	0.27	0.7	–	–	–
Control, 24 h	0.28	2.2	–	2.52	–
MA [a], μg/ml					
100	0.27	2.1	7	2.41	4
150	0.27	1.7	33	2.30	8
200	0.26	1.1	74	0.95	63
250	0.25	0.8	94	0.42	84
300	0.25	0.7	100	0.27	90
400	0.25	0.7	100	0.26	90

a   Mycophenolic acid (MA) was added to 48-h, washed, vegetative mycelia, incubated an additional 24 h

measured by ^{3}H-uridine incorporation. MA at levels of 250 μg/ml and above completely blocked PsV-F and viral dsRNA synthesis in the nonproliferating system. Under these nonproliferating conditions, the metabolite (MA) was most effective against the PsV-F viral system, especially when the virus replication is not linked to proliferating or primary phase growth conditions. Patulin and the other mycotoxic metabolites have not yet been tested in this nonproliferating assay system.

6 Conclusions

The definition of viruses in eukaryotic, filamentous fungi has been largely descriptive, because most reports have been ultrastructural observations. Biochemical and physiological studies on viruses associated with fungi have largely consisted of identification and purification of dsRNA components with limited studies of replication. Because the ubiquitous nature of dsRNA viruses in fungi is now well established, further comments will relate to how and why these dsRNA viruses may be accommodated and transmitted by the host mycelia. Transmission of the virus particles most certainly involves a cytoplasmic flow. The viruses are predominantly latent, and they possess simple capsid structures, exist in high titers in proliferating mycelia, and harbor an RNA replicase activity.

Obviously selective evolutionary pressures have resulted in these fungi for tolerance to a virus. A number of fungi are able to tolerate the presence of viral dsRNA at high cellular levels since dsRNA (fungal) has been demonstrated as a potent inhibitor of protein synthesis in mammalian systems.

How can one envisage the role of fungal metabolites in the control of virus replication in fungal tissue? In support of mycophenolic acid, this metabolite is produced in a nonsecondary fashion, i.e., synthesis parallels primary fungal growth. This metabolite is then present at early germination and outgrowth and could regulate a virus infection and/or replication. How other secondary metabolites with antiviral activity would control virus replication is difficult to understand, because the metabolites are produced after primary growth of the fungus.

The host organism in interacting with the virus replication must somehow exert nuclear control of the expression of these cytoplasmic dsRNA viruses. Studies with the yeast killer system have already indicated that cytoplasmic dsRNA and nuclear genes may act in concert (for literature, see Lemke, 1977).

Recent work with *P. stoloniferum* in our laboratory has shown that viruses are not always transmitted cytoplasmically in the conidiogenesis process. Approximately 7% of a random population of conidia contained (after outgrowth) no PsV-F particles. This lack of transmission of mature replicating virus particles to developing conidia might explain the variation in virus level or complete absence of detectable virus from single-conidia isolates. This inability to transmit may also be influenced by a nuclear gene expression in the maturing conidia. The single-conidia isolates that harbor no virus do not produce mycophenolic acid; therefore, these variants are not capable of supporting virus replication by a control mechanism not involving the production of an antiviral metabolite.

How mycophenolic acid and the other metabolites inhibit virus replication in vivo is not understood to date. Experiments with the PsV-F replicase in vitro indicate that both MA and patulin show no inhibition of replicase activity. Therefore, these metabolites appear to exhibit their biological activity against fungal viruses in a fashion other than binding directly to particle replicases. The inhibitory effect could be a preferential binding to dsRNA molecules, however, and could result in impairment to encapsidation by the viral proteins or translation of the information for capsid formation.

Considerable research is required to understand better the replication of these viruses and their tolerance by fungal hosts. Experimental infection to test the various models put forth by many workers is required if indeed these viruses are truly infectious.

Summary

Evidence from limited studies with fungal viruses indicates that these viruses generally are not involved in the production of fungal secondary metabolites. However, more recent studies have shown that a number of fungal metabolites exhibit inhibitory action against the replication in vivo of certain fungal viruses. Evidence is presented in support of a model by which selective mycotoxin-metabolites are capable of controlling or limiting the replication of viral dsRNA and/or viruses in proliferating fungi.

Information is presented on how specific metabolites such as mycophenolic acid, patulin, and gliotoxin interact with viruses and inhibit their replication in vivo. These studies were conducted with the viruses from *Penicillium stoloniferum*. The in vivo replication of PsV-F and PsV-S viruses in *P. stoloniferum* mycelia is discussed relative to fungal morphogenesis. Results are discussed in terms of how conidiogenesis may influence the transmission or persistence of these viruses during morphogenesis.

References

Abraham EP, Florey, HW (1949) Substances produced by Fungi Imperfecti and Ascomycetes. In: Florey HW, Chain E, Heatly NG, Jennings MH, Standards AG, Abraham, EP, Florey, ME (eds) Antibiotics, vol I. Oxford University Press, London, pp. 273–355

Ashoor SH, Chu FS (1973) Inhibition of alcohol and lactic dehydrogenases by patulin and penicillic acid in vitro. Food Cosmet Toxicol 11: 617–624

Atkinson N, Stanley NF (1943) Antibacterial substances produced by molds. The detection and occurrence of suppressors of penicillin activity. Aust J Exp Biol Med Sci 21: 249–253

Banks GT, Buck KW, Chain EB, Himmelweit F, Marks JE, Tyler JM, Hollings M, Last FT, Stone OM (1968) Viruses in fungi and interferon stimulation. Nature (London) 218: 542–545

Benigni R, Dogliotti E, Ignazzitto G, Volterra L (1978) Nucleic acid characterization of two virus-like particles in *Penicillium citrinum*. Mycovirus Newsl 6: 5–9

Birkinshaw JH (1965) Chemical constitutents of the fungal cell. In: Ainsworth GC, Sussman AS (eds), The fungi: An advaced treatise, vol 1. Academic Press, New York, pp 179–228

Borré E, Morgantini LE, Ortali V, Tonolo A (1971) Production of lytic plaques of viral origin in *Penicillium*. Nature (London) 229: 568–569

Bozarth RF (1972) Mycoviruses, A new dimension in microbiology. Environ Health Perspec 2: 23–39

Bozarth RF, Goenaga A (1977) Complex of virus-like particles containing double-stranded RNA from *Thielaviopsis basicola*. J Virol 24: 846–849

Day PR, Dodds JA, Elliston JE, Jaynes RA, Anagnostakis SL (1977) Double-stranded RNA in *Endothia parasitica*. Genetics 67: 1393–1396

Detroy RW, Still PE (1975) Fungal metabolites and viral replication in *Penicillium stoloniferum*. Dev Ind Microbiol 16: 145–151

Detroy RW, Freer SN, Fennell DI (1973) Relationship between the biosynthesis of virus-like particles and mycophenolic acid in *Penicillium stoloniferum* and *Penicillium brevi-compactum*. Can J Microbiol 19: 1459–1462

Dickens F, Jones HEH (1961) Carcinogenic activity of a series of reactive lactones and related substances. Br J Cancer 15: 85–90

Dickens F, Jones HEH (1965) Further studies on the carcinogenic action of certain lactones and related substances in the rat and mouse. Br J Cancer 19: 392–403

Dunkle LD (1974a) Double-stranded RNA mycovirus in *Periconia circinata*. Physiol Plant Pathol 4: 107–116

Dunkle LD (1974b) The relation of virus-like particles to toxin-producing fungi in corn and sorghum. Proc 28th Annu Corn Sorghum Res Conf, pp 72–81

Ellis LF, Kleinschmidt WJ (1967) Virus-like particles of a fraction of statolon, a mold product. Nature (London) 215: 649–650

Hollings M, Stone OM (1971) Viruses that infect fungi. Annu Rev Phytopathol 9: 93–118

Hooper IR, Anderson HW, Skell P, Carter HE (1944) The identity of clavacin with patulin. Science 99: 16–17

Jones, JB, Young JM (1968) Carcinogenicity of lactones. III. The reactions of unsaturated γ-lactones with L-cysteine. J Med Chem 11: 1176

Lange L, Olson LW (1978) Virus-like particles in *Synchytrium endobioticum* – the infectious agent of potato wart disease. Mycovirus Newsl 6: 14

Lapierre HD, Spire D (1974) Virus-like particles of phytopathogenic fungi. Abstr 1st Intersect Cong Int Assoc Microbiol Soc, p 72

Lapierre H, Astier-Manifacier S, Cornuet P (1971) Activité RNA polymerase associée aux preparations purifiées de virus du *Penicillium stoloniferum.* C R Acad Sci Ser D 273: 992–994

Lemaire JM, Jovan B, Perrator B, Sailly M (1971) Perspectives de lutte biologique contre les parasites des céréales d'origine tellurique en particular *Ophiobolus graminis* Sacc.Sci Agron Rennes, pp 1–8

Lemke PA (1977) Fungal viruses and agriculture. In: Virology in agriculture. Allanheld Osmun & Co, Montclair, New Jersey, pp 159–175

Lemke PA, Nash CH (1974) Fungal viruses. Bacteriol Rev 38: 29–56

Lemke PA, Ness TM (1970) Isolation and characterization of a double-stranded ribonucleic acid from *Penicillium chrysogenum.* J Virol 6: 813–819

Lemke PA, Nash CH, Pieper SW (1973) Lytic plaque formation and variation in virus titer among strains of *Penicillium chrysogenum.* J Gen Microbiol 76: 265–275

Loening UE (1967) The fractionation of high-molecular-weight ribonucleic acid by polyacrylamide-gel electrophoresis. Biochem J 102: 251–257

MacKenzie DE, Adler JP (1972) Virus-like particles in toxigenic Aspergilli. Abstr Am Soc Microbiol p 68

Misra A, Singh TKS, Yadar YP, Mishra B, Choudhary DP (1978) Mycovirus in *Helminthosporium.* Mycovirus Newsl 6: 14

Normansell ID, Holt G (1978) Viruses in strains of *Penicillum chrysogenum* impaired in penicillin biosynthesis. Mycovirus Newsl 6: 15

Norstadt FA, McCalla TM (1969) Patulin production by *Penicillum urticae* Bainier in batch culture. Appl Microbiol 17: 193

Rawlinson CJ, MacLean DJ (1973) Virus-like particles in axenic cultures of *Puccinia graminis tritici.* Tr Br Mycol Soc 61: 590–592

Rawlinson CJ, Hornby D, Pearson V, Carpenter JM (1973) Virus-like particles in the take-all fungus *Gaeumannomyces graminis.* Ann Appl Biol 74: 197–209

Rawlinson CJ, Muthyalu G, Deacon JW (1977) Natural transmission of viruses in *Gaeumannomyces* and *Phialophora* spp. Abstr 2nd Int Mycol Congr, p 558

Still PE, Detroy RW, Hesseltine CW (1975) *Penicillium stoloniferum* virus: Altered replication in ultraviolet-derived mutants. J Gen Virol 27: 275–281

Styer EL, Corbett HMK (1978) Intranuclear virus-like particles of *Phytophthora infestans* and *P. parasitica* var. *parasitica.* Mycovirus Newsl 6: 16–17

Volkoff O, Walters T, Dejardin RA (1972) An examination of *Penicillium notatum* for the presence of *Penicillium chrysogenum* type virus particles. Can J Microbiol 18: 1352–1353

Wood HA, Bozarth RF (1973) Purification of virus-like particles from an isolate of *Aspergillus flavus.* Abstr Am Soc Microbiol, p 198

Fungal Viruses and Killer Factors (Saccharomyces cerevisiae) *

M.H. VODKIN [1] and G.A. ALIANELL [2]

Department of Biology, University of South Carolina, Columbia, SC 29208/USA

1 Introduction

The purpose of this paper is to provide some details on the molecular biology of the killer factor of yeast and to discuss information subsequent to the last two reviews (Pietras and Bruenn, 1976; Wickner, 1976a). Certain strains of the yeast, *Saccharomyces cerevisiae,* are called killers because they secrete a toxin that at low pH kills other (sensitive) strains of the same species (Makower, 1963). Colonies can be tested for killing ability by replica plating on solid medium (pH 4.7), on which a confluent lawn of a sensitive indicator strain has been seeded. Killing will be visualized as a zone of clearing in the lawn surrounding the printed colony (Fig. 1).

Evidence will be presented suggesting that the genetic information for toxin is a double-stranded (ds) RNA that is encapsulated by a protein coat. It is thought that the protein coat is the translational product of a different dsRNA. Although the dsRNA itself is inherited in a non-Mendelian pattern there are many nuclear genes that influence its maintenance or expression.

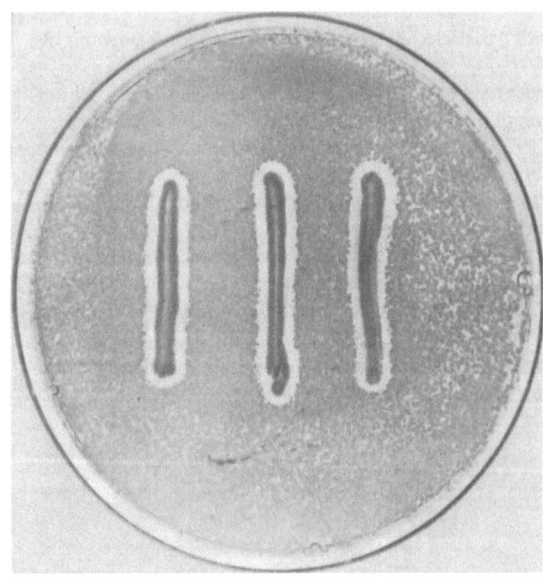

Fig. 1. Petri plate assay for killer strains. About 10^4 cells of a sensitive indicator strain were spread as a lawn on solid complete medium (pH 4.7). After drying, a pattern of killer cells was replica plated on it and the plate was incubated for 2 days at room temperature

* The study was supported by a grant to the first author from the National Institute of Health (GM 21438).

[1] Present address: Bldg 5, Room 324, NIH, Bethesda, MD, 20014/USA

[2] Present address: Department of Biophysical Sciences, University of Houston, Houston, Texas 77004/USA

2 Non-Mendelian Functions

Because of the resolving power of tetrad analysis, elements that show a non-Mendelian mode of inheritance can be easily detected in a single cross. Such an inheritance pattern has been found for mutations that affect toxin, resistance to toxin, or both. For purposes of notation, standard killer yeast strains are (K^+, R^+) to designate ability to kill and to resist the toxin, respectively.

2.1 Toxin

2.1.1 Neutral Strains

Neutral (K^-, R^+) is the phenotypic designation for a strain that does not kill but retains resistance or immunity to the toxin (Naumov, 1974, Somers and Bevan , 1968). It can be derived by mutation from a killer strain or can be found as such in a collection. When a neutral is crossed to a killer strain, the resulting diploid colonies are unstable and give mitotic segregants of killers and neutrals. Meiotic progeny of diploid killers include both killers and neutrals, reflecting the mixture of both kinds of cytoplasmic determinant. It is not known whether neutral strains synthesize an active toxin but fail to secrete it, synthesize a defective toxin, or fail to synthesize any product.

2.1.2 Weak Killers or Temperature-Sensitive Killers

Toxin activity is detectable at 30°C but not at 37°C. It shows much higher activity at 23°C and this temperature has routinely been utilized for visualization of toxin activity. There are mutants, however, that fail to excrete an active product at 30°C and show decreased activity at 20°C (Vodkin et al., 1974; Sweeney et al., 1976). In at least one case (unpublished data), the toxin itself showed an increased temperature sensitivity when compared to that of the wild-type killer parent strain.

2.1.3 Superkillers

Superkillers secrete more toxin or a more active toxin. They can be recognized on the standard Petri plate assay by a much wider zone of inhibition than wild-type killer strains at any temperature between 20°C and 30°C.

2.2 Resistance to Toxin

Strains (K^+, R^-) have been described that still kill but have lost resistance to toxin (Vodkin et al., 1974). These strains are conditionally lethal at low pH.

2.3 Toxin and Resistance to Toxin

Strains (K^-, R^-) that have lost both the ability to produce toxin and to resist it actually include a number of different classes which with two exceptions (suppressive nonkiller and diploid-dependent killer) are genetically indistinguishable but do contain altered patterns of dsRNA. They may also be distinguishable by the type of mutagenesis that induced them. The mutants can arise spontaneously at a low frequency (Fink and Styles, 1972) or can be induced by prolonged incubation at 37°C (Wickner, 1974), cycloheximide (Fink and Styles, 1972), or by standard mutagenesis protocols utilized for microorganisms. When crossed to a wild-type killer, the diploids are (K^+, R^+) and segregate 4:0 (K^+, R^+) to (K^-, R^-). The dsRNA alterations associated with some of these mutants will be discussed in Chapter 3.

2.3.1 Suppressive Nonkiller

In contrast to the genetic pattern for most classes of (K^-, R^-), the cytoplasm of suppressive nonkiller is dominant to that of wild-type killer (Somers, 1973; Vodkin et al., 1974; Sweeney et al., 1976). When crossed to a killer, many diploids are (K^-, R^-) or segregate as such and many tetrads show a 4:0 segregation of (K^-, R^-) to (K^+, R^+).

The induction of most suppressive nonkillers is through standard mutagenesis procedures. The incidence of such mutants as a class is generally quite low. In one particular genetic background, however, the frequency of suppressive nonkillers approached 70% of all nonkillers induced (Vodkin, 1977a). The reason for such a high frequency is not understood.

2.3.2 Diploid-Dependent Killers

Wickner (1976b) described mutants of the killer plasmid that showed a variable phenotype as a haploid: (K^-, R^+), (K^+, R^-), or (K^-, R^-). Merely crossing to any other haploid nonkiller, including one carrying a similar or identical cytoplasm, restored normal killing and resistance in the diploid. Sporulation of these diploids resulted in haploid spore clones that could show any of the three phenotypes previously mentioned, even within one tetrad. Thus, this mutation depends on a haploid genome or a single dose of a particular allele for its expression.

2.4 Relationship of Killer Plasmid to Other Elements with a Non-Mendelian Pattern of Inheritance

The yeast cell contains a number of elements that do not follow a Mendelian pattern of inheritance. They include rho (mitochondrial DNA), PSI, 2μ circular DNA, URE 3, and sporulation RNA. As far as can be determined the killer factor is not identical to any of these other elements. Strains that lack or have been cured of rho (Vodkin and Fink, 1973; Al-Aidroos et al., 1973), PSI (Lowenstein and Vodkin, unpublished data), 2μ circular DNA (Livingston, 1977), and URE 3 (Aigle and Lacroute, 1975) can still retain killing and resistance. Strains that have lost or lack the killer plasmid can still retain rho (Vodkin and Fink, 1973; Vodkin, 1977a), PSI (Vodkin, 1977a), 2μ circular DNA (Alianell and Vodkin, unpublished data), and sporulation RNA (Garvik and Haber, 1978). With a few exceptions nuclear mutations that affect the replication or expression of one element will not affect the replication or expression of another.

Because these elements are inherited in a non-Mendelian pattern, it is often assumed that they are localized in the cytoplasm. This assumption has been directly tested for the killer factor by Conde and Fink (1976). The killer factor could be transferred via heteroplasmon formation in the absence of nuclear fusion. Thus its heritability has been established as independent of the nucleus.

3 Double-Stranded RNAs Associated with Wild-Type and Mutated Killer Strains

3.1 Types of dsRNA Associated with Killer Strains

Standard killer strains usually contain two distinct species of dsRNA (Bevan et al., 1973; Vodkin et al., 1974). These RNAs have been called L and M or alternatively P1 and P2. They can be visualized and measured by gel electrophoresis, electron microscopy, or renaturation kinetics (Figs. 2, 3, and 4, respectively). L has a molecular weight of $2.7–3.0 \times 10^6$ daltons and accounts for 0.33% of total nucleic acids; M has

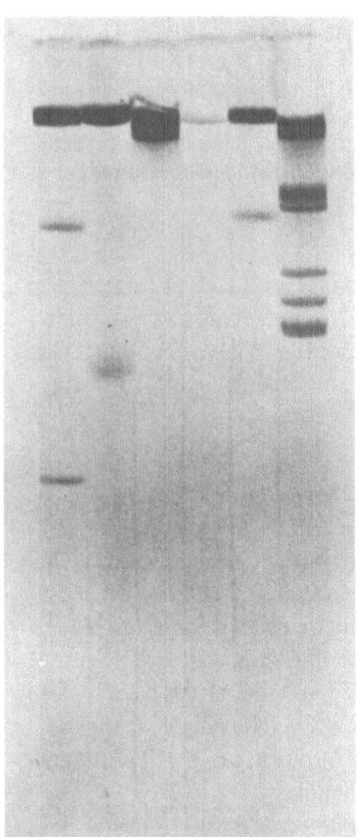

Fig. 2. Gel electrophorograms of dsRNA. RNA samples were submitted to electrophoresis on 5% polyacrylamide gels. The gels were then fixed and stained with methylene blue. Double-stranded reovirus has also been analyzed as a molecular weight marker. *From left to right,* immunity minus killer; suppressive nonkiller, cycloheximide-induced nonkiller, spontaneously derived nonkiller, and reovirus

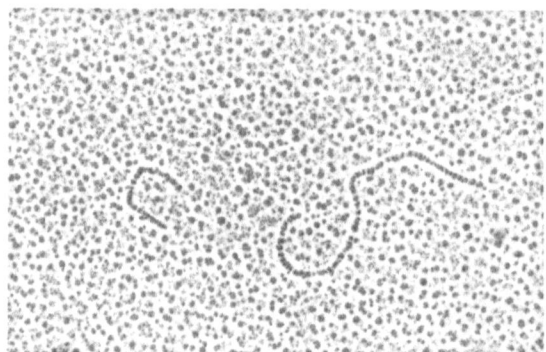

Fig. 3. Electron micrograph
of L and M dsRNA ($\times$ 40,000)

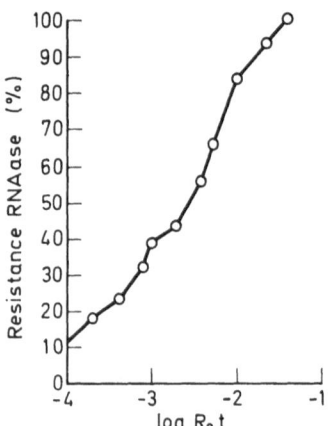

Fig. 4. Renaturation of L dsRNA. Forty ng of L
dsRNA, radiolabeled with ^{32}P (2.1 $\times$ 10^6 cpm/μg)
was mixed with 0, 15, 150, or 1500 ng of unlabeled L
dsRNA. The RNA was denatured, diluted, and allowed
to reanneal. At various time intervals, 50 μl aliquots
were removed and subjected to RNAase digestion in
high salt concentration

a molecular weight of 1.3–1.5 $\times$ 10^6 daltons and accounts for 0.01% of total nucleic
acids (Vodkin et al., 1974; Wickner and Leibowitz, 1976). Some mutants have additional bands with variable low molecular weights. In addition, Wickner and Leibowitz
(1976) have described XL dsRNA, which has a higher molecular weight than L and
may be its precursor or perhaps some type of replicative form.

3.2 Double-Stranded RNAs Associated with Mutants Derived from Killer Strains

Some mutants derived from wild-type killer strains that were discussed previously have
been associated with qualitative and/or quantitative alterations of dsRNA. The most
spectacular changes involve strains that are (K$^-$, R$^-$). All are deleted for M dsRNA
(Table 1). Depending on the mechanism of mutagenesis the amount of L dsRNA present in these strains also changes. Spontaneously derived mutants of this class typically
contain 1/3 the amount of L dsRNA present in (K$^+$, R$^+$). Cycloheximide and heat-
cured mutants have three times the amount of L dsRNA found in (K$^+$, R$^+$). Mutants
(K$^-$, R$^-$) derived by other mechanisms can fall into either class.

A very rare class has been reported that has lost all detectable dsRNA (Mitchell et
al., 1973; Vodkin, 1977a). It is not known why such strains occur so seldom and what

role the genetic background of the parent killer strain plays in the derivation. However, in the latter study almost 20% of the nonkillers fell into this class.

Some strains contain a dsRNA (S), which varies in molecular weight but is always less than that of M dsRNA. There are two types of pattern apparent. Killers that have lost resistance (K⁺, R⁻) contain L, M, and S dsRNA. Suppressive nonkillers lack M and

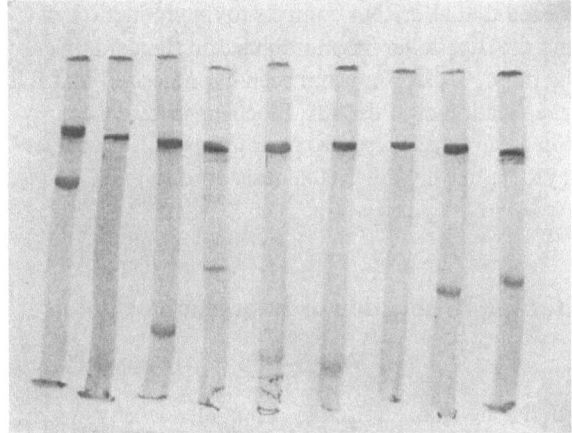

Fig. 5. Gel electrophorogram of suppressive nonkillers. *From left to right*, standard killer, cycloheximide-induced nonkiller, and various suppressive nonkillers

Table 1. The pattern of dsRNA in various types of yeast strains

	Phenotype	L	M	S	dsRNAs
	Killer (K⁺, R⁺)	+[a]	+	−[b]	
Plasmid mutations	Neutral (K⁻, R⁺)	+	+	−	
	Weak killer	+	+	−	
	Superkiller	+	++[c]	−	
	Immunity minus killer (K⁺, R⁻)	+	+	+	
	Spontaneous nonkiller (K⁻, R⁻)	+−[d]	−	−	
	Cycloheximide or heat-induced nonkiller (K⁻, R⁻)	++	−	−	
	Nonkillers derived from FM 11[e]	−	−	−	
	Suppressive nonkiller (K⁻, R⁻)	+	−	+	
Chromosomal mutations	*mak* (K⁻, R⁻)	+	−	−	
	kex (K⁻, R⁺)	+	+	−	
	rex (K⁺, R⁻)	+	+	−	
	ski[e] (K⁺, R⁺)	+	++	−	

[a] Presence of a dsRNA in the same amount as found in the standard killer

[b] Absence of dsRNA

[c] Presence of a dsRNA in significantly greater proportion than found in the standard killer

[d] Presence of a dsRNA in significantly less proportion than found in the standard killer

[e] See text

contain S dsRNA (Fig. 5). The different S dsRNAs are apparently derived from M by a large internal deletion with or without subsequent tandem duplication (Fried and Fink, 1977; Bruenn and Kane, 1978). Thus, S will form complex structures when viewed in the electron microscope after denaturation and self-annealing and will also hybridize to M dsRNA. However, L and M dsRNA do not share common sequences as judged by lack of cross-hybridization and common large T_1 oligonucleotides (Bruenn and Kane, 1978; Vodkin, unpublished observations).

The genetic evidence has suggested that M dsRNA controls toxin production, because all (K^+) strains contain it and the superkiller has more than the standard killer. Because no strain has M in the absence of L dsRNA, it has been hypothesized that L is necessary for the maintenance or replication of M dsRNA. Biochemical evidence, which will be considered in Section 5, supports these two hypotheses. It is probable that M dsRNA also controls the cytoplasmically inheritable resistance to toxin, though the mechanism is unknown.

4 Nuclear Genes Necessary for the Replication or Expression of the Killer Factor

The mutations mentioned in Section 2 are characterized by a non-Mendelian mode of inheritance. Additionally there are nuclear genes that affect either the maintenance of M dsRNA or the expression of the functions encoded by it (killing and resistance). No nuclear gene has been found to affect L dsRNA.

4.1 Maintenance of Killer Genes *(mak)*

At least 27 different complementation groups have been described which can directly or indirectly control the replication of M dsRNA (Wickner, 1978). Recessive mutations in any of these *mak* genes lead to a loss of M dsRNA and a strain that is (K^-, R^-).

The function of one *mak* gene has recently been determined to code for S-adenosylmethionine decarboxylase, one of the enzymes in the biosynthetic enzyme pathway for spermine and spermidine (Cohn et al., 1978). The functions of other *mak* genes have not been determined, but it is interesting to note that some drastically affect the biology of the yeast cell. For instance *pets* also does not maintain mitochondrial DNA and is temperature-sensitive for growth (Wickner and Leibowitz, 1976). An allele of *mak1* also acts as a temperature sensitive lethal mutation.

4.2 Superkiller Mutations

Toh-e et al. (1978) described four chromosomal mutations that enhance toxin activity of killer strains. (It is not known if resistance is "enhanced.") As in the case of the cytoplasmically inherited superkiller mutation, two of the *ski* mutants have increased amounts of M dsRNA. In *ski* backgrounds several killer plasmid mutations have been derived that require the *ski* mutation for maintenance and are lost in a wild-type background.

Mutations that Affect mak and ski Genes. A number of genes have been found that by-pass the requirement for *mak* genes. One appears to be a mitochondrial gene whose absence specifically bypasses *mak10–1* (Wickner, 1977). *KRB1* acts as a dominant chromosomal mutation that bypasses *mak7–1* and the *mak* defect of *pet18* mutants (Wickner and Leibowitz, 1977). Each *ski* mutant bypassed its own spectrum of different *mak* genes. To explain some of these bypass phenomena, Toh-e et al. (1978) have proposed that there are alternate pathways for the replication of M dsRNA.

4.3 Genes Affecting Toxin Production *(kex)* and Resistance to Toxin *(rex)*

These two classes of nuclear mutations affect only the expression of the killer plasmid and not its maintenance. The presence of L and M dsRNA has been demonstrated genetically and biochemically in such strains (Wickner, 1974; Wickner and Leibowitz, 1976). Leibowitz and Wickner (1976) noted that all *αkex2* strains are partially sterile, because they fail to excrete α-factor (pheromone) and do not respond to one of the A-factors. There is also a set of mutations which acts similarly to *kex* genes and have been classified as dominant *ochre* suppressions (Nesterova et al., 1975).

4.4 Killer Systems that Segregate in a Mendelian Pattern

There are two intriguing reports of a killer factor segregating as a chromosomal gene. Woods et al. (1974) noticed a spontaneously derived weakened killer that originated from a sensitive strain. The data are not conclusive but suggest a chromosomal pattern of segregation for this killer factor. More recently Naumov and Naumova (1978) also noted a killer strain that was seemingly under chromosomal control, and toxin production could not be cured by chemical agents that eliminated M dsRNA in other killer strains. However, it is unclear what types of dsRNA this particular strain has.

It should also be mentioned in connection with the above two strains that Shalitin and Fischer (1978) and Vodkin (1977b) have raised the possibility of transcripts of L and M dsRNA, respectively, that are integrated within the yeast DNA.

5 Products of dsRNA

The genetic evidence has indicated that M codes for toxin and resistance and that L codes for a function(s) necessary for the existence of M dsRNA. Biochemical evidence supports these hypotheses. Bostian et al. (1978) and Hopper et al. (1977) have translated the RNA's in vitro. L dsRNA codes for the major coat protein (Oliver et al., 1977) that encapsulates L and M in separate, but identical, particles. They estimated that L dsRNA had an additional capacity to code for another 80,000 daltons equivalent of protein. M dsRNA coded for a polypeptide that was precipitated by antibody specific for the toxin. The precipitated protein, however, was twice as large as the toxin produced in vivo. One can speculate that the residual portion of the in vitro protein is concerned with the mechanism of resistance.

Associated with the particles, though not necessarily encoded by dsRNA, are activities that replicate and transcribe dsRNA (Herring and Bevan, 1977; Brennan et al., 1978; Welsh and Leibowitz, 1978). The details of these processes are still being elaborated.

6 Other Killer Systems

Most of the data discussed thus far relate to laboratory strains of *Saccharomyces cerevisiae* all of which produce a functionally identical toxin and are resistant to that same toxin. It is possible that these laboratory strains were derived rather recently from a common ancestor. Other strains, such as brewer's yeast, sake yeast, and nonlaboratory strains, produce different types of toxins and patterns of resistance (Imamura et al., 1974; Rogers and Bevan, 1978; Young and Yagiu, 1978). The latter authors could also distinguish molecular weight differences among some of the M dsRNA's of independent origin. It will be interesting to see whether the rules elucidated for the original killer strains will be applicable to these new strains.

Summary

Most of the mycoviral genomes characterized to date consist of one or more segments of double-stranded (ds) RNA. The killer factor of baker's yeast, *Saccharomyces cerevisiae*, is an example of such a system. The standard laboratory strain killer contains two species of dsRNA with molecular weights estimated at 2.5×10^6 (L) and 1.1×10^6 (M) daltons, L dsRNA apparently codes for the three coat proteins that encapsidate either it or M as separate particles. M dsRNA controls toxin production and probably immunity of the host against toxin.

Cells that secrete the toxin are known as killers. The toxin shows maximum activity toward sensitive cells at pH 4.8. It has been purified to homogeneity and migrates as a single protein on SDS gels with a molecular weight of 11,000–15,000 daltons. The toxin binds first to cell walls of sensitive cells and then to the plasma membrane. Synthesis of DNA, RNA, and proteins ceases and ATP selectively leaks from the cells. ATP continues to be produced by these cells and energy production is required for cellular death. However, the exact mechanism of toxin action is not yet known.

Killer cells normally manifest a dual phenotype; secretion of toxin K^+ and resistance to toxin R^+. Mutations have been described that affect either or both of these phenotypes. Some of them act as either point mutations or aberrations of the dsRNA itself. The evidence for the above conclusions is based on the cytoplasmic mode of inheritance for the former class of mutants and in addition a gross molecular alteration of dsRNA for the latter class. There are K^- and K temperature sensitive that act as point mutations of M dsRNA. Complete deletions of M, which result in K^-R^- phenotype, can arise spontaneously at a low rate or be induced by mutagenesis; such mutations commonly lead to a threefold reduction of L dsRNA. Complete deletions of M along with a threefold increase in L dsRNA are induced at a high frequency either by elevated temperature or cycloheximide. Suppressive nonkillers, which are also K^-R^-, act as cytoplasmic dominants in crosses to K^+R^+ and lack M but have another dsRNA (S) with a variably lower molecular weight. Killers that have lost resistance, K^+R^-, contain L, M, and S dsRNA.

Almost all strains of baker's yeast contain some L dsRNA. There had been an earlier report that 5-fluorouracil could cure this dsRNA. Mutagenizing strains that happened to contain *psi*, another extrachromosomal element, resulted in a high frequency of cells lacking all detectable dsRNA. These strains will serve as useful controls for investigating various parameters and functions of yeast dsRNA.

A number of nuclear mutations are also known to affect the maintenance or expression of the killer factor. They are designated as *mak, kex,* or *rex.* Thus far maintenance genes have been described only for M dsRNA. A mutation of one of these *mak* genes results in a concomitant loss of M dsRNA, killing, and resistance. About 30 independent genes have been mapped and some regulate other biological processes, including replication of mitochondrial DNA, mating, and sporulation. Mutations of *kex* genes prevent synthesis or secretion of active toxin. Mutations of *rex* genes render killer cells sensitive to toxin. Neither of the latter two classes of mutation alters the dsRNA.

As in other mycoviruses, a virion-associated RNA polymerase has been described. Presumably the replication of dsRNA of killer factor is catalzyed in vivo by this enzyme. Despite this mode of replication, there are suggestions that all or part of the nucleotide sequence of L dsRNA also resides in DNA of the host yeast cell. The evidence is based on hybridization of L dsRNA to host DNA. The resultant RNA–DNA hybrid can be visualized in a Cs_2SO_4 gradient and is sensitive to digestion by RNAse H. The function of these sequences in the host DNA is not understood.

In conclusion, the killer factor of yeast is a segmented dsRNA genome that depends on many host functions. There are questions as to what relationship it has to other fungal viruses and why it depends on so many host genes.

References

Adler J, Wood HA, Bozarth RF (1976) Virus-like particles from killer, neutral, and sensitive strains of *Saccharomyces cerevisiae.* J Virol 17:472–476

Aigle M, Lacroute F (1975) A nonmitochondrial, cytoplasmically inherited mutation in yeast. Mol Gen Genet 136:327–335

Al-Aidroos K, Somers JM, Bussey H (1973) Retention of cytoplasmic killer determinants in yeast cells after removal of mitochondrial DNA by ethidium bromide. Mol Gen Genet 122:323–330

Bevan EA, Herring AJ, Mitchell DJ (1973) Preliminary characterization of two species of dsRNA in yeast and their relationship to the "killer character." Nature (London) 245:81–86

Bostian KA, Hopper JE, Rogers DT, Tipper DJ (1978) Translational analysis of the dsRNA genome of the killer-associated viruslike particles of *Saccharomyces cerevisiae.* M-ds RNA encodes toxin polypeptide. In: 9th Int Conf Yeast Genet Mol Biol, p 103, University of Rochester

Brennan V, Hastie N, Bruenn J (1978) Characterization of yeast viral transcriptase. In: 9th Int Conf Yeast Genet Mol Biol, p 102, University of Rochester

Bruenn J, Kane W (1978) Relatedness of the double-stranded RNAs present in yeast virus-like particles. J Virol 26:762–772

Cohn MS, Tabor CW, Tabor H, Wickner RB (1978) Spermidine or Spermine requirement for killer double-stranded RNA plasmid replication in yeast. J Biol Chem 253:5225–5227

Conde J, Fink GR (1976) A mutant of *Saccharomyces cerevisiae* defective for nuclear fusion. Proc Nat Acad Sci USA 73:3651–3655

Fink GR, Styles CA (1972) Curing of a killer factor in *Saccharomyces cerevisiae.* Proc Nat Acad Sci USA 69:2846–2849

Fried HM, Fink GR (1977) Electronmicroscope heteroduplex analysis of yeast killer double-stranded RNA plasmids. In: Sherman F, Broach J (eds) The molecular biology of yeast. Cold Spring Harbor Laboratory, Cold Spring Harbor, p 43

Garvik B, Haber JE (1978) New cytoplasmic genetic element that controls 20 S RNA during sporulation in yeast. J Bacteriol 134:261–269

Herring AJ, Bevan EA (1977) Yeast virus-like particles possess a capsid-associated single-stranded RNA polymerase. Nature (London) 268:464–466

Hopper JE, Bostian KA, Rowe LB, Tipper DJ (1977) Translation of the L species ds RNA genome of the killer-associated viruslike particles of *Saccharomyces cerevisiae.* J Biol Chem 252:9010–9017

Imamura T, Kawamoto M, Takaoka Y (1974) Characteristics of main mash infected by killer yeast in sake brewing and the nature of its killer factor. J Ferment Technol 52:293–299

Leibowitz MJ, Wickner RB (1976) A chromosomal gene required for killer plasmid repression, mating, and sporulation in *Saccharomyces cerevisiae*. Proc Natl Acad Sci USA 73:2061–2065

Livingston DM (1977) Inheritance of the 2 μm DNA plasmid from *Saccharomyces*. Genetics 86: 73–84

Makower RB (1963) The inheritance of a killer character in yeast *(Saccharomyces cerevisiae)*. Proc XI Int Congr Genet 1:202

Mitchell DJ, Bevan EA, Herring AJ (1973) The correlation between ds RNA in yeast and the "killer character." Heredity 31:133–134

Naumov GI (1974) Comparative genetics of yeast. XIV. Analysis of wire strains of *Saccharomyces* neutral to the killer strain type K2. Genetika 10:130–136

Naumov GI, Naumova TI (1978) Comparative genetics of yeast. XVII. A new type of killer in *Saccharomyces* yeast. Genetika 14:138–144

Nesterova GF, Zekhnov AM, Inge-Vechtonov SG (1975) Dominant nonsense suppressions suppressing the antagonistic activity of the yeast *Saccharomyces cerevisiae*. Genetika 11:96–103

Oliver SG, McCready SJ, Holm C, Sutherland PA, McLaughlin CS, Cox BS (1977) Biochemical and physiological studies of the yeast virus-like particle. J Bacteriol 130:1303–1309

Pietras DF, Bruenn, JA (1976) The molecular biology of yeast killer factor. Int J Biochem 1:173–179

Rogers D, Bevan EA (1978) Group classification of killer yeasts based on cross-reactions between strains of different species and origins. J Gen Microbiol 105:199–202

Shalitin C, Fischer I (1975) Abundant species of poly (A)-containing RNA from *Saccharomyces cerevisiae*. Biochim Biophys Acta 414:263–272

Somers JM (1973) Isolation of suppressive mutants from killer and neutral strains of *Saccharomyces cerevisiae*. Genetics 74:571–579

Somers JM, Bevan EA (1968) The inheritance of the killer character in yeast. Genet Res 13:71–83

Sweeney TK, Tate A, Fink GR (1976) A study of the transmission and structure of double-stranded RNAs associated with the killer phenomenon in *Saccharomyces cerevisiae*. Genetics 84: 27–41

Toh-e A, Guerry P, Wickner RB (1978) Chromosomal superkiller mutations and killer plasmid mutations affecting plasmid maintenance. In: 9th Int Conf Yeast Genet Mol Biol, p 103, University of Rochester

Vodkin MH (1977a) Induction of yeast killer factor mutations. J Bacteriol 132:346–348

Vodkin M (1977b) Homology between double-stranded RNA and nuclear DNA of yeast. J Virol 21:516–521

Vodkin MH, Fink GR (1973) A nucleic acid associated with a killer strain of yeast. Proc Nat Acad Sci USA 70:1069–1072

Vodkin MH, Katterman F, Fink GR (1974) Yeast killer mutants with altered double-stranded ribonucleic acids. J Bacteriol 117:681–686

Welsh JD, Leibowitz MJ (1978) Particular DNA-independent RNA polymerase from killer yeast. In: 9th Int Conf Yeast Genet Mol Biol, p 101. University of Rochester

Wickner RB (1974) "Killer character" of *Saccharomyces cerevisiae:* curing by growth at elevated temperatures. J Bacteriol 117:1356–1357

Wickner RB (1976a) Killer of *Saccharomyces cerevisiae:* a double-stranded ribonucleic acid plasmid. Bacteriol Rev 40:757–773

Wickner RB (1976b) Mutants of the killer plasmid of *Saccharomyces cerevisiae* dependent on chromosomal diploidy for expression and maintenance. Genetics 82:273–285

Wickner RB (1977) Deletion of mitochondrial DNA bypassing a chromosomal gene needed for maintenance of the killer plasmid of yeast. Genetics 87:441–452

Wickner RB (1978) Twenty-six chromosomal genes needed to maintain the killer double-stranded RNA plasmid of *Saccharomyces cerevisiae*. Genetics 88:419–425

Wickner RB, Leibowitz MJ (1976) Chromosomal genes essential for replication of a double-stranded RNA plasmid of *Saccharomyces cerevisiae:* the killer character of yeast. J Mol Biol 105: 427–434

Wickner RB, Leiboowitz MJ (1977) Dominant chromosomal mutation bypassing chromosomal genes needed for killer RNA plasmid replication in yeast. Genetics 87:453–469

Woods DR, Bevan EA (1968) Studies on the nature of the killer factor produced by *Saccharomyces cerevisiae.* J Gen Microbiol 51:115–126

Woods DR, Ross IW, Hendry DA (1974) A new killer factor produced by a killer/sensitive yeast strain. J Gen Microbiol 81:285–289

Young TW, Yagiu M (1978) A comparison of the killer character in different yeasts and its classification. Antonie van Leeuwenhoek 44:59–77

Fungal Viruses and Killer Factors – Ustilago maydis Killer Proteins*

Y. KOLTIN and R. LEVINE

Department of Microbiology, Faculty of Life Sciences, University of Tel-Aviv, Ramat Aviv/Israel

1 Introduction

In studies on the genetics and biochemistry of the host–virus interactions, as in many other biological phenomena, the preferred systems are those with a short generation time in which biochemistry and genetics can be studied simultaneously. The fungi, as a relatively simple group of eukaryotic organisms, have played a prominent role in basic research on the genetic expression at the cellular level (Beadle and Tatum, 1941; Beadle, 1946) as a model system that is amenable to both genetic and biochemical analysis. However, in all the research on virus–host interactions the fungi stand out as a unique group in which little reference to viruses was ever made. Numerous extrachromosomally inherited phenomena in fungi have been reported, such as unstable somatic segregation (Arlett et al., 1962; Grindle, 1964; Jinks, 1966), senescence (Marcou and Schecroun, 1969), vegetative death (Jinks, 1959), and even plaque formation (Koltin et al., 1973). However, the presence of viruses in the fungi was shown clearly only in the early 1960's.

It is clear today that viruslike particles are not uncommon in the fungi (Lemke and Nash, 1974; Lemke, 1976), yet the expression of their presence in cells is known in very few systems of which the better characterized are baker's yeast (Somers and Bevan, 1969; Wickner, 1976) and corn smut *Ustilago maydis* (Puhalla, 1968; Day and Dodds, in press). In both cases the expression of the presence of their viruses is by the excretion of a substance that leads to growth inhibition or death of sensitive cells. This killer effect is an intragenus phenomenon, the yeast cells are sensitive to the yeast substance and cells of *Ustilago maydis* are sensitive only to the substance excreted by *Ustilago* cells.

2 The Killer Phenomenon and a Double-Stranded RNA Virus

In *Ustilago maydis* cumulative data from various studies with over 150 monosporidial cultures from North America, Mexico, and Poland revealed four strains (one of which was eventually lost) that displayed the ability to inhibit the growth of other strains of the same species (Puhalla, 1968 and unpublished; Anagnostakis, unpublished; Koltin and Day, unpublished) as well as closely related species (Koltin and Day, 1975). The

---

* The study was supported in part by a grant to the first author from the Branch for Basic Research, Israel Academy of Sciences.

special strains were found only among the North American isolates. This inhibitory phenomenon is referred to as the killer phenomenon although, as will be shown, some killer strains affect cell viability, while others cause lethality of most cells and the prolongation of the generation time among the survivors.

The killer phenomenon was shown by genetic tests to be related to some cytoplasmic entity and inherited as a cytoplasmic trait. By formal crosses between killer strains and sensitive strains it was shown that the majority of the progeny from such crosses (70%–100%) are of the killer phenotype and tetrads, the meiotic products of single zygotes that are expected to display a 1:1 ratio for each trait determined by a nuclear gene, are phenotypically uniform either as killers or nonkillers (Puhalla, 1968; Koltin and Day, 1976a). Furthermore, in classical tests for the transmission of cytoplasmic traits, namely, heterokaryon transfer tests (Jinks, 1964), the transmission of the killer phenotype by cytoplasmic mixing was demonstrated for all known killer strains (Day and Anagnostakis, 1973; Koltin and Day, 1976a).

The killer phenotype was correlated with the presence of a viruslike particle, 41 nm in diameter containing double-stranded RNA (dsRNA) in all killer cells and the absence of such particles in sensitive cells (Wood and Bozarth, 1973; Koltin and Day, 1976a). Although the relation of the virus particles to the killer phenomenon is unquestioned today, deviations from these correlations were demonstrated originally by Koltin and Day (1976b) and later by Koltin (1977) and Koltin and Kandel (1978). It is currently accepted that virus particles can be harbored by sensitive strains as well as killer strains and it is the genetic information contained in the virus particles that determines the expression of killer activity and not the mere presence of dsRNA or viral coats.

3 The Killer Specificity

The interrelations between the killer strains indicate mutual inhibition among the three killers isolated from the natural population. Therefore, the killer phenomenon in *Ustilago* is viewed as a phenotypically similar phenomenon displayed by few strains with a high degree of specificity. The three original killer strains are each sensitive to the other and thus their specificities were designated P1, P4, and P6. The substance produced by these killers is designated KP1, KP4, and KP6, accordingly.

In addition to the killing property by which the specificity can be determined, other criteria such as affinity of the killer substances to concanavaline A, effect of pH on the synthesis of the killer substance and thermolability of each type of killer substance can all differentiate between the three inhibitory substances (Kandel and Koltin, 1978). KP1 and KP6 bind irreversibly to con-A, KP4 can be eluted from con-A by 1M NaCl; KP1 is produced in media that maintain neutrality for sufficient time, KP4 and KP6 are less dependent on the pH of the medium and produce the killer substance within a range of pH from pH4 to pH9; KP1 is very sensitive to heating, whereas KP4 and KP6 can withstand even 5–7 min at 80°C with a loss of no more than 50% of the activity. Most significant, however, for the distinction between the three substances is the genetic resistance to the killer substances determined by three different nuclear genes (Puhalla, 1968; Koltin and Day, 1976a) (Fig. 1). The nuclear genes for resistance,

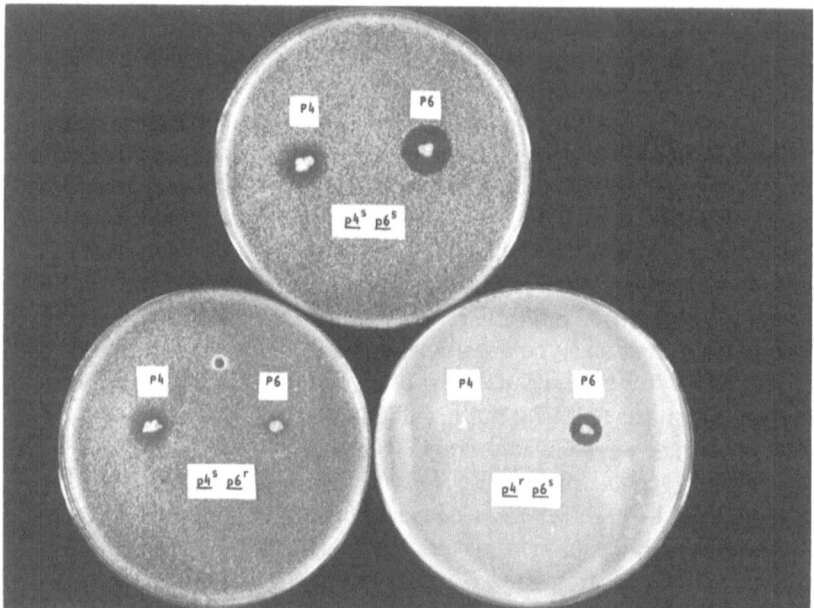

Fig. 1. Resistance of sensitive strains to killer toxins KP4 and KP6 determined by nuclear genes. Killer strains P4 *(left)* and P6 *(right)* spotted on lawns of sensitive strains carrying the sensitive alleles to both killer types *(upper center)*, sensitive allele to P4 and resistance to P6 *(lower left)* and the resistance allele to P4 and sensitivity to P6 *(bottom right)*

as judged in diploids, are recessive and the allele for sensitivity is dominant. These relations between the alleles suggest that resistance represents the loss of receptors for the killer substance and that each killer substance with a different specificity binds to a different receptor.

The kinetics of growth in the presence of the killer substance (Table 1) as well as the killing kinetics both provide additonal criteria for the distinction between the substances excreted by different killer strains. The results obtained in studies of the growth kinetics in the presence of the toxin can be seen qualitatively in the zones of inhibition

Table 1. Growth kinetics of sensitive cells in the presence of killer toxins

	Time (h)			
Killer toxin	24	48	72	120
–	2,600,00[a]	80,000.00	170,000.00	110,000.00
KP1	0.03	0.02	2.40	7.50
KP4	0.07	0.08	0.48	20.00
KP6	0.00	0.00	0.00	0.00

[a] $N/N_0 - N_0$, 10^3 cells inoculated at t = 0, N-cell titer at the specified time

around the killer strains (Fig. 1). The zone of inhibition around killers of type P1 and P4 lacks a sharp boundary and within the zone of inhibition a diffuse background is always noticed. The zone of inhibition of the P6 killer is always sharp and the background is clear even if kept under growth conditions for a few days. The zones of inhibition of P1 and P4 diminish gradually if kept for a few days in growth conditions. Quantitative determinations of cell density of the sensitive cells grown in the presence of the killer substance of each type clarify the situation with respect to the characteristics of the zones of inhibition of the three killer types. KP6 kills the sensitive cells immediately even in incubation periods of 2 min. KP1 and KP4 seem to have a dual effect, killing over 99% of the sensitive cells and prolonging the generation time of the remaining cells. The survivors are not resistant to KP1 and KP4. The nature of this dual effect is unclear but is repeatable with highly concentrated preparations of the substance from P1 and P4.

4 Nature of the Killer Substance

The killer substance of P1 was shown to be nondialyzable and sensitive to proteolytic enzymes (Hankin and Puhalla, 1971). The same was found to be true for the substances from P4 and P6. None of the killer substances is sensitive to nucleases and all are precipitable with ammonium sulphate. Therefore, the inhibitory substances are thought to be proteins.

Gel electrophoresis of the killer proteins revealed additional subtle differences between the three proteins related to charge and conformation of the molecules (Davis, 1964; Ornstein, 1964). The position of the toxins was determined by elution of the activity from the gels and by staining of replicate gels with Coomassie Blue. Both KP1 and KP6 were eluted from 7.5% polyacrylamide gels (pH 8.3) but KP1 migrated much faster than KP6. The protein of KP4 could not be eluted from the gels as were KP1 and KP6. However, KP4 and KP6 were eluted from pH 4.3 gels but KP1 could not be detected in the same conditions.

Once the killer proteins are denatured with SDS all three killer proteins migrate as a single band to the same position. All three toxins can be eluted from the SDS gels in an active form suggesting that the denaturation caused by heat and SDS before electrophoresis is reversible, a property that is anticipated if the molecule is relatively small. The molecular weight of the three killer proteins as determined, in 12.5% polyacrylamide gels containing SDS with a series of markers including insulin (MW 5,800) as the smallest marker, is ca. 10,000.

Two of the killer substances have been purified by a procedure involving acetone precipitation of the proteins and gel filtration with a series of polyacrylamide beads. Following this purification procedure KP4 and KP6 can be eluted, from 12.5% gels containing SDS, from a single position and Coomassie Blue does not reveal any additional protein bands on these gels. The results from gel electrophoresis suggest that the active protein consists of one subunit. This does not exclude, however, the possibility that the active form of the toxin is a homomultimer. Based on the in vitro complementation tests between the inactive proteins from various mutants complementation was obtained in some cases. These results can be interpreted as resulting from intermolecular complementation by the formation of homomultimers involving defective monomers of each mutant affected at a different site.

Amino acid analysis of only KP6 has been completed thus far and the results are in good agreement with the determination of the molecular weight by SDS gel electrophoresis. The protein contains 82 amino acids corresponding to a molecular weight of ca. 10,300. Aspartic acid and glutamic acid constitute almost 30% of the amino acids. The basic amino acids constitute less than 10% of the protein. The degree of variation between the proteins, which may also provide some clue to the specificity of the proteins, will become apparent with the completion of the amino acid analysis of KP1 and KP4.

5 On the Mode of Action of Killer Proteins

The mode of action of the killer proteins is as yet unknown. Furthermore, it is unclear whether the killer phenomenon is not a byproduct of some primary function of these proteins. The selective advantage of killer strains is unclear and many selective disadvantages can be indicated. If sexuality is of prime importance for the existence of the species it is disadvantageous. Killer strains in nature are not common and resistant strains also are not very frequent. Therefore, the killer function, if active in nature, would operate in the population as an isolation mechanism. For *Ustilago maydis* as a plant pathogen dependent for pathogenicity on heterokaryosis (Burnett, 1975) the killer function seems disadvantageous. The advantage in diversion of energy for the production of the killer proteins is also unclear since the killer proteins as antibiotics were shown to be ineffective against 25 species belonging to 12 genera of bacteria and even against competitive fungal species (Koltin and Day, 1975). The killer proteins are only effective against other Ustilaginales, but these are so specialized physiologically that under natural conditions the different species have no common host.

Some experimental data shed the greatest doubt on the significance of the killer function as a function of some selective advantage that would favor this type of virus–host interaction. Crosses can be performed between killer and sensitive cells. The crosses are performed on seedlings and these develop neoplastic growth as a result of the infection. Attempts to recover killer activity from galls infected with compatible killer strains with identical killer specificity have all failed (Puhalla, unpublished; Koltin, unpublished). This result suggests adsorption of the killer activity on some component of the plant tissue since neither heterokaryons nor diploid cells cease to produce the killer protein. The adsorption properties of the killer proteins to some Millipore filters suggest a very strong affinity to various polymers of cellulose. One is led to conclude that the killer phenomenon is a reflection of some cellular activity which can be identified in the laboratory as killer activity.

Information pertaining to the mode of action of the toxins was indicated from two directions of experimentation. In attempts to expand the killing specificity of individual killer strains, crosses were performed between the different killer strains in all possible combinations. The results indicated some restrictions on the inclusion of different viral genomes in a single cell. The outcome of such crosses defined specific exclusion relations between the three different killers in which P4 excludes unilaterally P6 and partially P1, P1 and P6 mutually exclude each other (Koltin and Day, 1976a). These studies did not indicate, however, any relation of the exclusion function to the killer activity. Such indications were obtained in studies with mutants of the virus with incomplete genomes. In an attempt to map viral functions the exclusion function was studied with mutants of both P4 and P6 that contain only part of the dsRNA segments known to occur in each type of virus (Koltin, 1977; Koltin and Kandel, 1978; Koltin et al., in press). The results of these studies pointed out the specific segments of dsRNA of the viral genome essential for the killer function. Deletion mutants lacking these molecules were also devoid of the capacity to display the exclusion relations known in their progenitors. Furthermore, the loss of the exclusion function permitted for the first time the introduction of segments of both P4 and P6 within a single cell with the restoration of killer function of P6 upon the introduction of the specific molecules associated with P6 killer activity. These results suggest that the toxin is associated with some nucleolytic activity perhaps similar to the restriction enzymes (Meselson et al., 1972) but operating on dsRNA in vivo.

It is clear today that 10%–20% of the sensitive strains contain dsRNA that corresponds to part of the viral genome associated with the different killers. These data could suggest some type of cellular defence mechanism by the toxin such as protection against superinfection or infection by other RNA viruses. Nonetheless, the involvement of the information for killer expression in the exclusion relations directed the attention to a search for nuclease activity associated with the purified killer toxin in in vitro studies as an indication of its in vivo function.

The activity of the killer toxins of both P4 and P6 was tested on various nucleic acid substrates and in its effect on in vitro protein synthesis. Inhibition of protein synthesis in the reticulocyte system (Table 2) and in other systems was demonstrated. The nuclease activity associated with the purified toxins degrades effectively both the 28S and 18S subunits of the ribosomal RNA (Fig. 2). Furthermore, the dissociation of polysomes to predominantly monosomes is noticed following brief incubation of polysomes with the killer proteins (Fig. 3). The specificity of these reactions has not been determined as yet. In the in vitro tests the activity is not restricted to ssRNA substrates

Table 2. Inhibition of in vitro protein synthesis (reticulocyte system) by killer proteins

Killer protein	Concentration $M \times 10^{-5}$	Incorporation of C^{14}-leucine (cpm)	Inhibition %
–	0.00	7000	0
4	0.75	2599	63
	1.50	2227	68
	3.00	1393	80
	3.75	938	87
6	3.50	1071	85
	4.37	1104	84
	8.75	621	91
	17.50	413	94

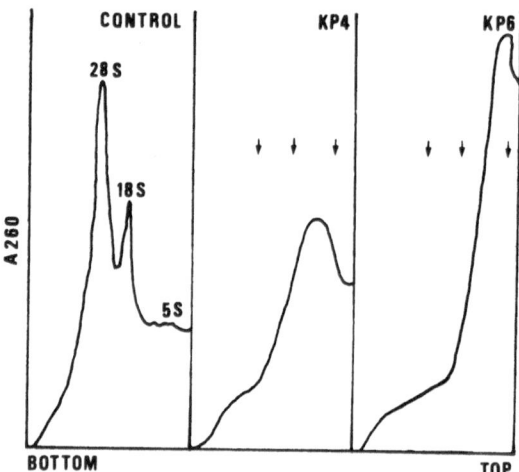

Fig. 2. Sucrose gradients (8%–18%) of ribosomal RNA after incubation (30 min, 25°C) with killer protein type 4 (KP4) and type 6 (KP6). 1.5×10^{-5} M of KP4 and 2.2×10^{-5} M of KP6 were used. Incubation was performed in 2×10^{-2} M Tris (pH 7.4) containing 5×10^{-2} M KCl and 0.4×10^{-2} M $MgCl_2$

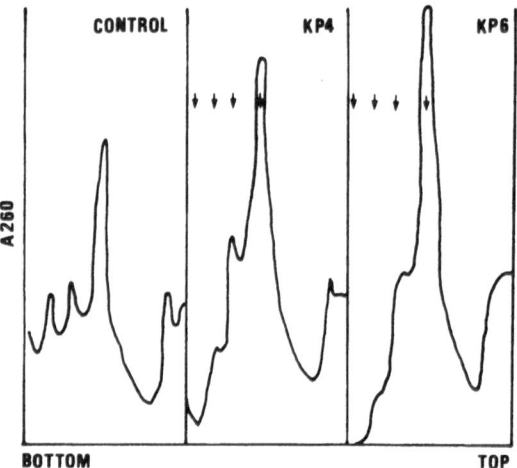

Fig. 3. Sucrose gradients (15%–40%) of polysomes after incubation (7.5 min, 37°C) with KP4 (0.18×10^{-5} M) and KP6 (0.87×10^{-5} M). *Arrows* indicate (from *left* to *right*) position of polysomes containing 4, 3, 2, and 1 ribosome(s)

but acts also on dsRNA molecules of both the *Ustilago* virus and Reovirus. Thus, it may conceivably attack such molecules in the cell.

In the exclusion mechanism a high degree of specificity should be anticipated especially since some of the molecules of all three types of virus seem to be identical and the size of major coat-associated protein of all three viruses is identical, suggesting some shared information by these viruses of the various types (Koltin et al., in press). However, this specificity may not be detected in vitro. Nevertheless, it has been established that some nuclease activity is associated with the killer toxins of *Ustilago* and its toxic effect may function in a manner resembling the colicins of bacteriophages of type E2 and E3 (Hardy, 1975; Holland, 1975). An additional interpretation of the findings is that the toxin is bifunctional, acting as a very specific nuclease in the exclusion phenomenon in the cell and as a nonspecific nuclease as a killer toxin once excreted from the cell. As small as it may seem, internal processing and interactions with other components may serve to modify the activity of the toxin once excreted from the cell. Studies with nonkiller mutants of both P4 and P6 suggest that 2 dsRNA molecules are involved in killer expression one of which is large enough to code for the killer protein, the other very small and possibly involved only in some type of processing of the toxin. Current experiments are directed toward the determination of the substrate specificity of the nuclease and the identification of the in vivo precursor of the killer toxin.

Summary

Among the population of *U. maydis* few strains display the ability to kill or inhibit the growth of other strains. The inhibitory substances are proteins. Three such proteins were identified, one from each of three killer strains. Each killer is resistant (immune) to its protein and sensitive to the proteins excreted by the other killer strains. The three proteins were purified by column chromatography on DEAE Sephadex, P30 and P60. The molecular weight of each protein is ca. 10,000 as determined by SDS gels and confirmed for one of the proteins by amino acid analysis. The proteins do not cross-react and differ by various criteria including binding affinity, pH dependence, and thermal sensitivity. Also, different nuclear genes determine the resistance to each of the three proteins. Each killer specificity is correlated with a unique dsRNA genome of otherwise indistinguishable VLPs found in killer strains. The diameter of the particles and the size of the coat protein in all three types is identical. The mapping of viral functions by deletion mapping and molecular replacement of specific dsRNA molecules indicate a similarity in position of the killer information. Thus far, all the spontaneous and induced mutations leading to the loss of killing affected only the viral genome. Nuclear mutations affecting killer expression and maintenance were not detected. The overriding of the viral exclusion relations was achieved by deletion of specific dsRNA molecules. New hybrid viral genomes were synthesized. These results provide the first indication on the mode of action of the killer proteins, suggesting an involvement in an exclusion-type phenomenon. However, the role of the killer proteins is not viewed as one of a major selective advantage and it is suggested that a more dramatic role may be played by the dsRNA as suppressive agents. DsRNA molecules are not rare in strains of *U. maydis* and *Saccharomyces cerevisiae* and in both cases are known to suppress the replication of the normal killer. The killer character is found in very few strains and is limited in its action to closely related species.

References

Arlett CF, Grindle M, Jinks JL (1962) The "red" cytoplasmic variant of *Aspergillus nidulans*. Heredity 17:197–209

Beadle GW (1946) Genes and the chemistry of the organism. Am Sci 34:31–53

Beadle GW, Tatum EL (1941) Genetic control of biochemical reactions in *Neurospora*. Proc Natl Acad Sci USA 27:499–506

Burnett JH (1975) Mycogenetics. John Wiley and Sons, London

Davis BJ (1964) Disc electrophoresis. II. Method and application to human serum proteins. Ann NY Acad Sci 121:404–427

Day PR, Anagnostakis SL (1973) The killer system in *Ustilago maydis:* Heterokaryon transfer and loss of determinants. Phytopathology 63:1017–1018

Day PR, Dodds JA (1979) Viruses of plant pathogenic fungi. In: Lemke PA (ed) Viruses and plasmids in fungi. Dekker, New York

Grindle M (1964) Nucleo-cytoplasmic interactions in the "red" cytoplasmic variant of *Aspergillus nidulans*. Heredity 19:75–95

Hankin L, Puhalla JE (1971) Nature of a factor causing interstrain lethality in *Ustilago maydis*. Phytopathology 61:50–53

Hardy KG (1975) Colicinogeny and related phenomena. Bacteriol Rev 39:464–515

Holland IB (1975) Physiology of colicin action. Adv Microbiol Physiol 12:55–139

Jinks JL (1959) Lethal suppressive cytoplasm in aged clones of *Aspergillus glaucus*. J Gen Microbiol 21:397–409

Jinks JL (1964) Extrachromosomal inheritance. Prentice-Hall, Inc, Englewood Cliffs, New Jersey

Jinks JL (1966) In: Ainsworth GL, Sussman AS (eds) The fungi, vol II, pp 619–660. Academic Press, New York

Kandel J, Koltin Y (1978) Killer phenomenon in *Ustilago maydis*: comparison of the killer proteins. Exp Mycol 2:270–278

Koltin Y (1977) Virus-like particles in *Ustilago maydis*: mutants with partial genomes. Genetics 86: 527–534

Koltin Y, Day PR (1975) Specificity of *Ustilago maydis* killer proteins. Appl Microbiol 30:694–696

Koltin Y, Day PR (1976a) Inheritance of killer phenotypes and double stranded RNA in *Ustilago maydis*. Proc Natl Acad Sci USA 73:594–598

Koltin Y, Day PR (1976b) Suppression of the killer phenotype in *Ustilago maydis*. Genetics 82: 629–637

Koltin Y, Kandel JS (1978) Killer phenomenon in *Ustilago maydis*: the organization of the viral genome. Genetics 88:267–276

Koltin Y, Berick R, Stamberg J, Ben-Shaul Y (1973) Virus-like particles and cytoplasmic inheritance of plaques in a higher fungus. Nature New Biol 241:108–109

Koltin Y, Mayer I, Steinlauf R (in press) Killer phenomenon in *Ustilago maydis*. Mapping viral functions. Mol Gen Genetics

Lemke PA (1976) Viruses of eukaryotic microorganisms. Annu Rev Microbiol 30:105–145

Lemke PA, Nash CH (1974) Fungal viruses. Bacteriol Rev 38:29–56

Marcou D, Schecroun J (1959) La sénescence chez *Podospora anserina* pourrait être due à des particules cytoplasmiques infectantes. C R Acad Sci (Paris) Ser-D 248:280–283

Meselson M, Yuan R, Heywood J (1972) Restriction and modification of DNA. Annu Rev Biochem 41:442–466

Ornstein L (1964) Disc electrophoresis. I. Background and theory. Ann NY Acad Sci 121:321–349

Puhalla JE (1968) Compatibility reactions on solid medium and interstrain inhibition in *Ustilago maydis*. Genetics 60:461–475

Somers JM, Bevan EA (1969) The inheritance of the killer character in yeast. Genet Res (Camb) 13:71–83

Wickner RB (1976) Killer of *Saccharomyces cerevisiae*: a double stranded ribonucleic acid plasmid. Bacteriol Rev 40:757–773

Wood HA, Bozarth RF (1973) Heterokaryon transfer of virus-like particles associated with a cytoplasmically inherited determinant in *Ustilago maydis*. Phytopathology 63:1019–1021

Methods

Screening for Viruses in Human Pathogenic Fungi

J.P. ADLER

Biological Sciences Department, California State Polytechnic University, Pomona,
CA 91768/USA

1 Introduction

Many fungal plant pathogens have been examined for the presence of viruses in order
to study the effects that they may have on their fungal hosts (Lemke, 1977). Only a
few studies have focused attention on fungi such as *Histoplasma capsulatum* (Rose and
Adler, 1977), *Torulopsis glabrata, Cryptococcus neoformans, Candida albicans* and
other *Candida* sp. (Kandel and Stern, 1979; Stumm et al., 1977) which cause disease
in humans. The viral screening techniques used to study fungal human pathogens are
very similar to those employed for the plant pathogens, except that special precautions
must be taken in culturing and harvesting these organisms because of their potential
pathogenicity for the investigators.

  Those fungi which grow primarily as yeast cells both at 25°C and 37°C on common
laboratory media such as Sabouraud's (Sab) and Brain Heart Infusion (BHI) agar are
often considered opportunistic human pathogens because they usually cause serious
infections only in compromised hosts. Dimorphic fungi, on the other hand, which
grow in the yeast phase at 37°C on enriched media, such as BHI, and as sporulating
mycelium at 25°C on Sab are very pathogenic for both healthy and debilitated individuals (Rippon, 1974). *Histoplasma capsulatum* which causes the disease histoplas-
mosis, is an example of the latter type of fungus. This fungus has been examined for
double-stranded ribonucleic acid (dsRNA) and virus presence by the reporting investi-
gator and will be used in this discussion to describe the various methods that can be
used to study viruses in fungal human pathogens.

2 Materials and Methods

2.1 Culturing

Eight strains of *H. capsulatum* were grown in both the yeast and mycelial phases. All
inoculations had to be done in a Class III laminar flow hood with surgical masks and
gloves.

  When the organism was inoculated into 250 ml Erlenmeyer flasks containing 25 ml
of Sab broth and incubated at 25°C, it grew as a mycelium which sporulated after two
to three months. The yield was 20 to 40 g wet weight per liter of broth. The maximum
possible yield was desired since each experiment required a minimum of 5 to 10 g of
tissue.

Conversion of the organism to the yeast phase was done by using BHI test tube slants to which one ml of sterile human or bovine serum had been added. Pieces of the mycelium were placed on the slant and in the serum, tubes were sealed with masking tape and then incubated at 37°C for 10 to 14 days. Sealing the tubes prevented drying out of the biphasic medium during this time. After incubation, most of the fungus was growing in the yeast phase, but there were varying amounts of nonsporulating mycelium present, depending on the strain being studied. The mass culturing of the yeast cells was obtained by growing the converted organism in 25 X 150 mm test tubes on BHI slants to which one ml of BHI broth had been added. The large test tubes were needed to provide the maximum surface area for growth of the organism. The cultures were incubated at 37°C for 10 to 21 days. This technique yielded 20 to 30 g wet weight of yeast cells per liter of medium.

2.2 Harvesting

Viable spores can easily infect individuals who are exposed to aerosols in which they are present. Therefore, each flask of sporulating mycelium was incubated for 48 h with 50 ml of 70% ethanol to kill the fungus prior to harvesting the mycelium. The mycelium was collected in the hood with a Buchner funnel lined with cheesecloth, weighed, and then stored in the freezer.

The yeast cells were scraped off the surface of the agar slants, resuspended in water, pelleted by centrifugation in tared bottles and also stored in the freezer.

2.3 Extraction

Both the yeast and mycelial tissue were homogenized for 1 to 3 min in a Bronwill mechanical cell homogenizer. The buffers used included a Na_2HPO_4 (0.03 M), NaH_2PO_4 (0.03 M), NaCl (0.15 M), pH 7.5 buffer for the yeast cells and a KH_2PO_4 (0.1 M), K_2HPO_4 (0.1 M), pH 7.0 buffer for the mycelium. Previous work in this area has shown that different buffers are required for maximum viral yield in different fungal-virus systems (Adler et al., 1976). The cell-free extracts were subjected to a low speed (12,500 g) and then a high speed (130,000 g) centrifugation. Following another low speed centrifugation, some of the resuspended pellets were layered onto 10% to 40% sucrose density gradients and centrifuged at 133,000 g for 1.0 h at 4°C. The gradients were fractionated with an ISCO density gradient fractionator and analyzer. A single-phase SDS-phenol procedure (Diener and Schneider, 1968) was used to extract the nucleic acid from other resuspended pellets.

2.4 Detection of dsRNA

Following a 1-h incubation at 25°C with 5 μg/ml of DNAse, the nucleic acid extracts from both the yeast and the mycelial phases of eight strains of *H. capsulatum* were placed in 0.75% agarose diffusion plates and tested against anti-polyinosinic-polycytidylic acid (polyIC) antibodies (Rose and Adler, 1977). These extracts were also mixed with polyacrylamide microbeads coupled with anti-polyIC antibodies to determine by the indirect agglutination test if the extracts contained dsRNA (Ohlson et al., 1977). The rocket immunoelectrophoresis test was also used to detect dsRNA in some of these extracts (Adler and Del Vecchio, in press).

Pelleted sucrose density gradient absorption peaks were treated with 10% sodium lauryl sulfate for 30 min at 60°C (Ratti and Buck, 1972), layered onto polyacrylamide-agarose disc gels (2.5%) and electrophoresed for 5 h at 6V/cm with a boric acid (0.9 M), TRIS (0.9 M), EDTA (0.0223 M) buffer, pH 8.3 diluted 1:10. To detect nucleic acid, the gels were stained for 12 to 24 h with 0.01% aqueous toluidine blue and then destained with water (Herring and Bevan, 1974).

2.5 Detection of Virus

Carbon-coated Formvar grids were floated on the resuspended pellets obtained from the differential centrifugation of cell-free extracts and material from sucrose density gradient absorption peaks. These samples were negatively stained with aqueous 2% sodium phosphotungstate and examined in a Zeiss EM9 electron microscope.

3 Results

When the initial mycelial cultures from eight strains of *H. capsulatum* were subcultured in Sab broth and tested for dsRNA presence by the Ouchterlony immunodiffusion test (Fig. 1), the indirect agglutination test (Fig. 2), and in some cases the rocket immuno-electrophoresis test (Fig. 3), only two of the eight strains gave positive results. When the mycelial cultures were converted to the yeast phase and subcultured on BHI slants with BHI broth, four other strains were found to contain dsRNA. It was observed that no strain contained dsRNA in both phases of growth, and one strain contained no detectable dsRNA.

Cell-free extracts from those four strains which contained dsRNA in the yeast phase were subjected to sucrose density gradient centrifugation. The largest absorption peaks

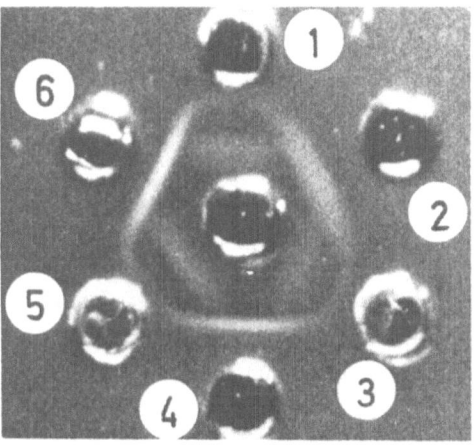

Fig. 1. Ouchterlony immunodiffusion test showing presence of dsRNA in a cell-free extract of *Histoplasma capsulatum* (strain No. 28). *Center well* contains anti-polyinosinic-polycytidylic antisera; wells *2, 4,* and *6* contain polyinosinic-polycytidylic acid and wells *1, 3,* and *5* contain the cell-free extract from strain No. 28

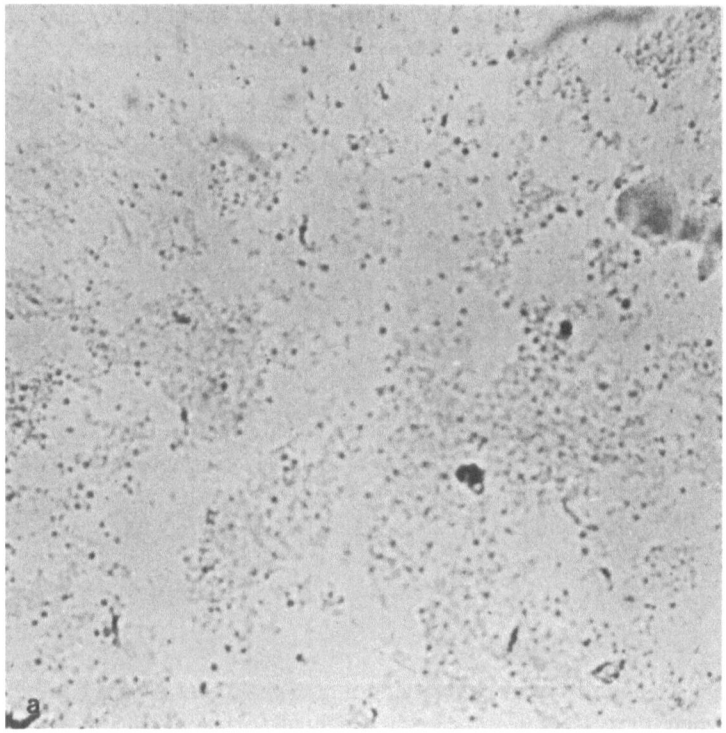

Fig. 2.a Indirect agglutination of polyacrylamide beads coupled with anti-polyinosinic-polycytidylic gamma globulin when mixed with phenol-extracted dsRNA from *Histoplasma capsulatum* (strain No. 27)

were obtained with strain No. 28 (Fig. 4a) and spherical virus particles were seen in all three peaks when they were examined in the electron microscope (Fig. 5). The viruses in each peak ranged from 40 nm to 130 nm in diameter, but the majority were between 60 and 66 nm in diameter. They were more variable in size than other fungal viruses (Lemke and Nash, 1974). The results demonstrated that sucrose density gradient centrifugation did not separate these different size particles, and that they were particularly susceptible to sodium phosphotungstate since they broke down when stained for more than 30 s.

The sedimentation coefficients of the virus peaks in Figure 4 (108S, 168S, and 227S, respectively) were estimated by comparing the absorption profile in Figure 4a with the sucrose density gradient profile for *Saccharomyces cerevisiae* cell-free extracts similarly treated (Fig. 4b) (Adler et al., 1976).

Preliminary studies with polyacrylamide-agarose disc gel electrophoresis indicated that the first peak (108S) contained empty particles and the other two peaks (168S and 227S) contained dsRNA, i.e., only the latter two peaks contained material which stained pink with toluidine blue.

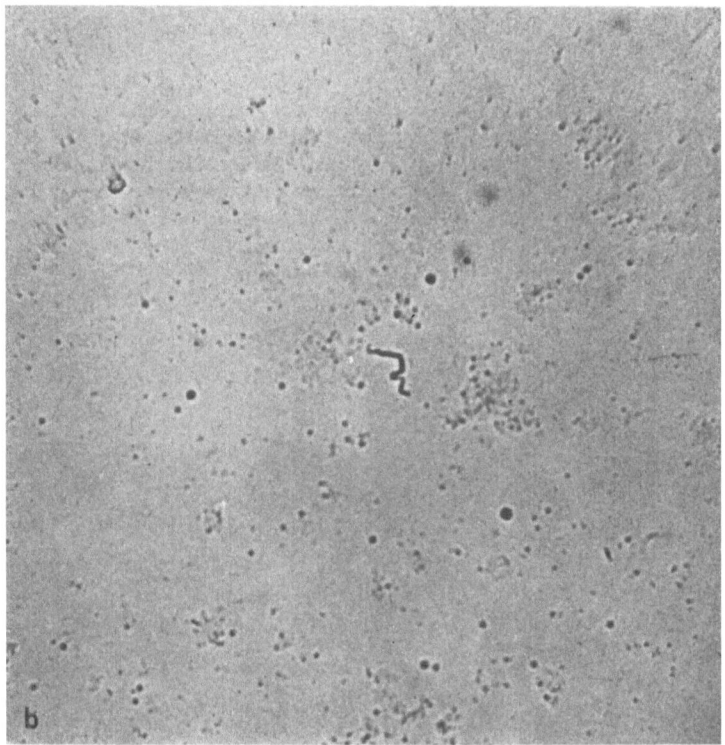

Fig. 2. b No agglutination of the beads when mixed with Tris-KCl buffer (0.1 M, pH 7.5)

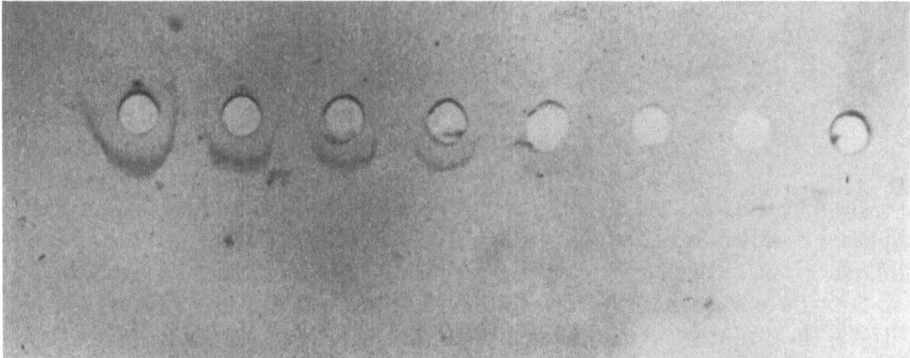

Fig. 3. Rocket immunoelectrophoresis of polyinosinic-polycytidylic acid into a 1% agarose gel containing anti-polyinosinic-polycytidylic gamma globulin stained with Coomassie Brilliant Blue R. Wells *from right to left* contain decreasing concentrations of polyinosinic-polycytidylic acid: 50, 25, 12.5, 5.25, 3.125, 1.56, 0.78 and 0.39 μg

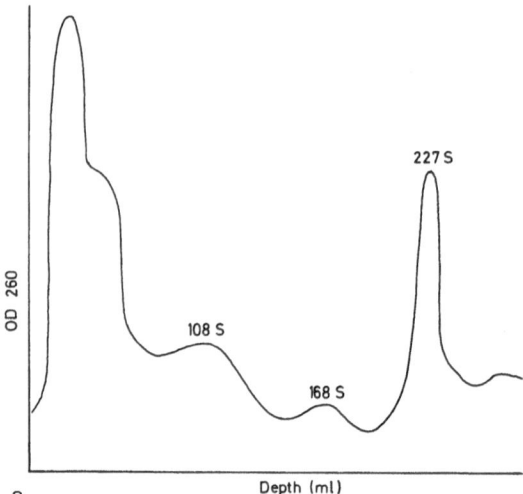

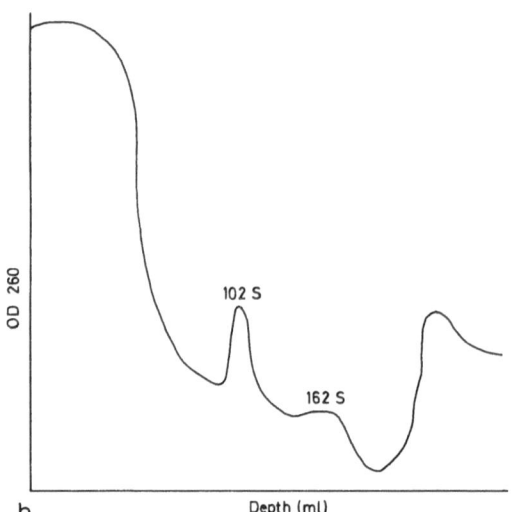

Fig. 4. a Sucrose density gradient profile of a cell-free extract of 9.5 g wet weight of *Histoplasma capsulatum* (Strain No. 28) grown at 37°C on a biphasic BHI medium. Sedimentation coefficients of the particles (108S, 168S, and 227S) in the absorption peaks were estimated by comparison with the sedimentation coefficients of the viruses extracted from *Saccharomyces cerevisiae* (sensitive strain) (102S and 162S) similarly centrifuged (b)

When the virus-containing yeast phase of strain No. 28 was converted to the mycelial phase by incubation of the yeast cells in Sab broth at 25°C, for the first time, the virus particles were detectable in mycelial extracts of this strain. The virus particles had the same sedimentation coefficients as the particles extracted from the yeast phase (Fig. 6). These particles were maintained in the fungal tissue following conversion of the virus-infected mycelium to the yeast phase again.

Alternatively, when the virus-infected yeast phase of strain No. 28 was incubated on BHI slants with BHI broth at 25°C, rather than 37°C, the concentration of virus particles/g wet weight was markedly decreased (Fig. 7). One other study showed that unlike strain No. 28, the dsRNA which was present in the yeast phase of strain No. 27, was not detectable when the fungus was converted to the mycelial phase.

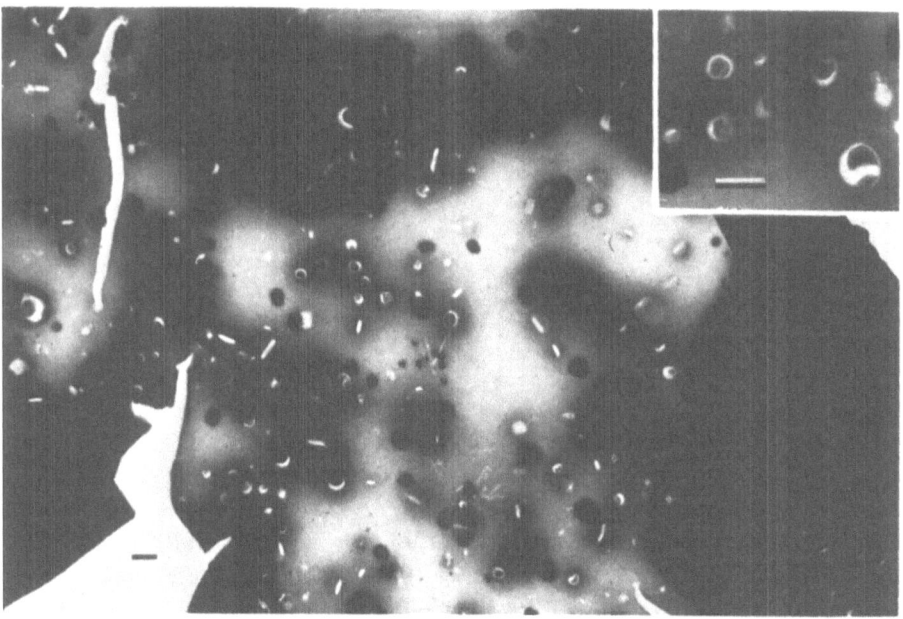

Fig. 5. Spherical virus particles from the sucrose density gradient absorption peaks of *Histoplasma capsulatum* (strain No. 28) stained with 2% sodium phosphotungstate; bar = 100 nm. *Inset:* same material at higher magnification; bar = 100 nm

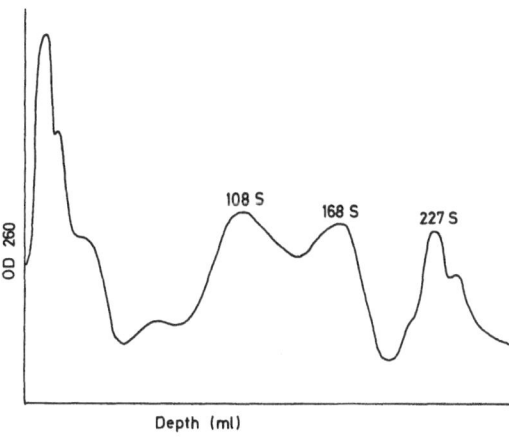

Fig. 6. Sucrose density gradient profile of a cell-free extract of 3 g wet weight of *Histoplasma capsulatum* (strain No. 28) grown at 25°C in Sabouraud's broth. Sedimentation coefficients of the particles in the absorption peaks were estimated to be 108S, 168S, and 227S

4 Conclusion

Double-stranded RNA and spherical virus particles were detected in some strains of *H. capsulatum.* The virus particles were not all uniform in size, but the majority were 60 to 66 nm in diameter. Because these particles were disrupted by exposure to sodium phosphotungstate, the results indicated that other stains should be used to visualize

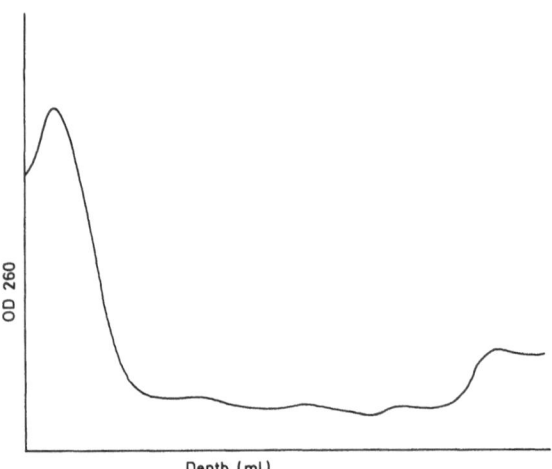

Fig. 7. Sucrose density gradient profile of a cell-free extract of 9.5 g wet weight of *Histoplasma capsulatum* (strain No. 28) grown at 25°C on a biphasic BHI medium

them. The estimated sedimentation coefficients of the virus particles (108S, 168S, and 227S) suggested that, like other fungal viruses, the *H. capsulatum* viruses were multi-component (Wood, 1973).

Some of the results reported here suggest that viral expression may be closely linked to host deoxyribonucleic acid (DNA) information. This possibility has been discussed in relation to other fungal virus systems: for example, (1) the presence of high concentrations of virus particles in *Aspergillus flavus* in the absence of cytoplasmic dsRNA and viral nucleic acid (Wood et al., 1974); (2) isolation of a lytic agent from two species of *Candida* following UV irradiation (Mehta and Shirodkar, 1977); (3) the presence of an agent in *Coprinus congregatus* that has the characteristics of an integrated virus in the dikaryon and a cytoplasmic virus in the monokaryon (Ross, 1977); and (4) conversion of virulent, dsRNA-free, single-spore isolates (derived from hypovirulent dsRNA-containing strains of *Endothia parasitica*) to hypovirulent dsRNA-containing strains after UV irradiation (Day and Dodds, 1979). In *H. capsulatum,* although ds-RNA could not be isolated from the initial mycelial subcultures of strain No. 28, ds-RNA and virus particles could be isolated from the mycelium if the fungus was first converted to the yeast phase by growing it on BHI at 37°C. This phenomenon may be a characteristic of only certain strains and it may be only a temporary condition in a given strain. One study also suggested that temperature may affect virus expression since very few particles could be detected in the fungal tissue when it was grown on BHI at 25°C, rather than at 37°C. These observations suggest that further investigation of some of these possible host—virus relationships is needed to understand better the role of fungal viruses in nature.

Summary

Eight strains of the dimorphic fungus *Histoplasma capsulatum* were examined for double-stranded ribonucleic acid (dsRNA) and virus presence in both the yeast and mycelial phases of growth. The mycelia were grown in Sabouraud's broth at 25°C, and the yeast cells were cultured on a biphasic medium of Brain Heart Infusion (BHI) agar and BHI broth or serum. Following disruption in the Bronwill mechanical cell homogenizer, the material from both the mycelia and the yeast cells was subjected to a series of differential centrifugations. Only two of the eight strains contained dsRNA in the cell-free extracts prepared from the initial subcultures of the mycelial phase. When the mycelial cultures were converted to the yeast phase, dsRNA was detected in four other strains. Cell-free extracts from the yeast phase of these four strains were layered onto 10% to 40% sucrose density gradients and centrifuged at 133,000 g for 1.0 h at 4°C. These preparations contained spherical virus particles of various sizes (40 to 130 nm in diameter) which sedimented at 108S, 168S and 227S. Although dsRNA could not be isolated from the initial mycelial subcultures of one strain, virus particles could be isolated from the mycelium if the fungus was first converted to the yeast phase. Temperature may also affect virus expression in this strain since the virus particle concentration was markedly decreased when it was grown on BHI at 25°C, rather than at 37°C. These results suggested that viral expression may be closely linked to host deoxyribonucleic acid information.

References

Adler J, Wood HA, Bozarth RF (1976) Virus-like particles from killer, neutral and sensitive strains of *Saccharomyces cerevisiae*. J Virol 17:472–476

Adler JP, Del Vecchio VG (1979) Specialized assays for detection of fungal viruses and double-stranded RNA. In: Lemke PA (ed) Viruses and plasmids in fungi. Dekker, New York

Day PR, Dodds JA (1979) Viruses of plant pathogenic fungi. In: Lemke PA (ed) Viruses and plasmids in fungi. Dekker, New York

Diener TO, Schneider IR (1968) Virus degradation and nucleic acid release in single-phase phenol systems. Arch Biochem Biophys 124:401–412

Herring AJ, Bevan, EA (1974) Virus-like particles associated with the double-stranded RNA species found in killer and sensitive strains of the yeast *Saccharomyces cerevisiae*. J Gen Virol 22:387–394

Kandel JS, Stern TA (1979) The killer phenomenon in pathogenic yeast. Antimicrob Agents and Chemother 15:568–572

Lemke PA (1977) Fungal viruses and agriculture. In: Virology in agriculture. Allanheld Osmun and Co., pp 159–175

Lemke PA, Nash CH (1974) Fungal viruses. Bact Rev 38:29–56

Mehta JP, Shirodkar MVN (1977) Lytic agent from *Candida* species. Mycovirus Newslett 5:11

Ohlson GB, Adler JP, Rose S (1977) Screening for doublestranded RNA by indirect agglutination. Abs 2nd Int Mycol Cong, Tampa, p 488

Ratti G, Buck KW (1972) Virus particles in *Aspergillus foetidus:* a multicomponent system. J Gen Virol 14:165–175

Rippon JW (1974) Medical mycology: the pathogenic fungi and the pathogenic actinomycetes. Saunders, New York

Rose SA, Adler JP (1977) The presence of dsRNA in the dimorphic fungus, *Histoplasma capsulatum*. Abs 2nd Int Mycol Cong, Tampa, p 575

Ross IK (1977) An infectious disorder for meiosis in *Coprinus*. Abs 2nd Int Mycol Cong, Tampa, p 577

Stumm C, Hermans JMH, Middlebeck EJ, Croes AF, DeVries GJML (1977) Killer-sensitive relationships in yeast from natural habitats. Antonie van Leuvenhoek 43:125–128

Wood HA (1973) Viruses with double-stranded RNA genomes. J Gen Virol 20:61–85

Wood HA, Bozarth RF, Adler J, Mackenzie DW (1974) Proteinaceous virus-like particles from an isolate of *Aspergillus flavus*. J Virol 13:532–534

Immunochemical Detection of Viruslike Particles and Double-Stranded RNA from Agaricus bisporus

V.G. DEL VECCHIO, C. DIXON, J. FRITZINGER, and P.A. LEMKE

Mellon Institute, Carnegie-Mellon University, Pittsburgh, PA 15213/USA

1 Introduction

The mushroom virus disease of *Agaricus bisporus* is associated with the presence of double-stranded RNA (dsRNA) and several viruslike particles (VLP), two spherical with diameters of 25 nm and 34 nm and one bacilliform with overall dimensions of 19 × 50 nm (Hollings, 1962; Dieleman-van Zaayen, 1972; Saksena, 1975; Lemke, 1976; and Fig. 1). Antisera to these VLP and to dsRNA have been prepared. An investigation of the VLP has been carried out by immune electron microscopy, counter-immunoelectrophoresis, immunoelectrophoresis, and the enzyme-linked immunosorbent assay (ELISA). Immunochemical profiles of the VLP and of dsRNA from various mushroom tissues have been determined in an attempt to relate virus titers to disease expression.

Immune electron microscopy (IEM) is an extremely sensitive assay for viruses in extracts of infected plant and animal material (Milne and Luisoni, 1977). The value of IEM lies not only in its simplicity but also in that it can be qualitatively used to distinguish specific serotypes of viruses.

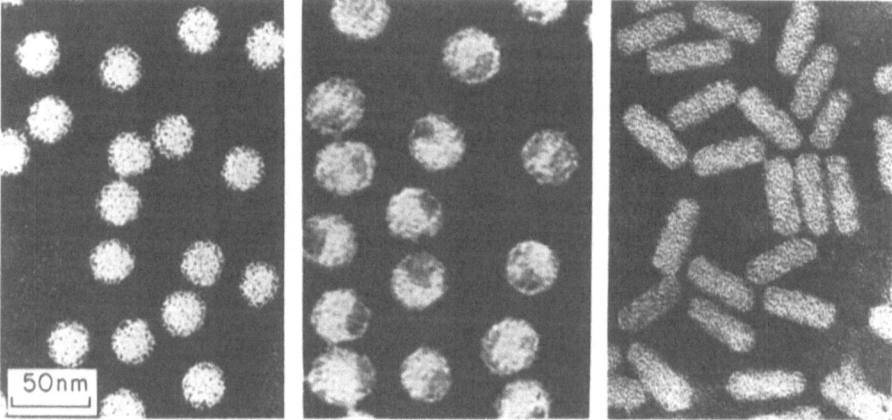

Fig. 1. Electron micrograph of purified VLP of *Agaricus bisporus* examined without use of serologically specific anti-VLP serum. [Reproduced, with permission, from the Ann. Rev. Microbiol. *30*, 105–145 (1976) by Annual Reviews Inc.]

A variety of IEM techniques was applied to detect VLP associated with a disease of the cultivated mushroom. Antisera to spherical VLP, 25 nm and 34 nm in diameter, as well as antisera to synthetic dsRNA were used to test for the presence of these virus particles in both mushrooms and vegetative cells (Del Vecchio et al., 1978). Both 25 nm and 34 nm particles have been found at low titer in symptomless cultures.

In addition to the IEM techniques, counter-immunoelectrophoresis (CIEP) was employed for the screening of the viruses of *Agaricus bisporus*. CIEP involves the electrophoretic migration of antigens with the simultaneous electroendosmotic flow of antibodies. The resulting precipitates indicate immune recognition. The CIEP test described herein is specific for the presence of 25 nm and 19 × 50 nm, bacilliform, VLP.

Two further assays, immunoelectrophoresis (IE) and enzyme-linked immunosorbent assay (ELISA) were conducted, using mainly antiserum prepared to 25 nm particles. The IE procedure has established serological heterogeneity among 25 nm particles. The increased sensitivity of the ELISA technique has confirmed the presence of 25 nm particles in diseased as well as symptomless mushrooms and vegetative cells.

2 Immune Electron Microscopy

Crude extracts of virus-infected mushrooms or vegetative cells were prepared by grinding tissue with dry ice in a mortar and pestle. An equal amount (w/v) of phosphate buffer (pH 7.2, 0.05 M) was added to the broken cells. The resulting suspension was

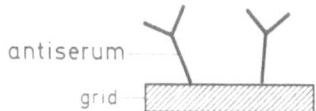

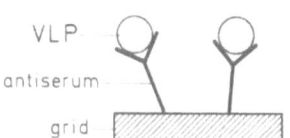

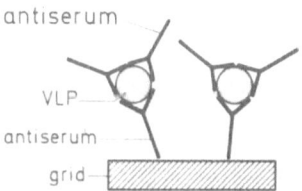

Fig. 2. Procedure for IEM: *(top)* diluted (1/100) antiserum specific for the VLP (anti-VLP) or antiserum to synthetic dsRNA (antipoly I:poly C) was placed on a Formvar-coated copper grid (400 mesh) for 5 min; *(middle)* following washing with 20 drops of phosphate buffer, one drop of the crude extract was placed on the grid for 15 min, and the grid was washed with phosphate buffer; *(bottom)* recognition of VLP by electron microscopy was enhanced through decoration by placing a drop of the diluted antiserum on the already trapped virus particles. Following 5 min incubation with the added antiserum, the grid was washed with phosphate buffer, distilled water, and negatively stained with 1% PTA. All grids were examined on a Phillips EM 200

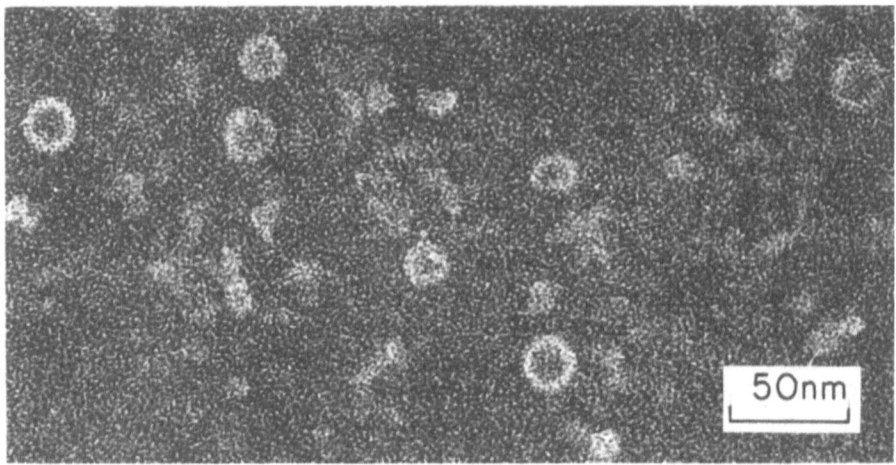

Fig. 3. Conventional electron micrograph (without the use of serologically specific anti-VLP serum) of the fraction used in Figure 4b, but without dilution. [Reproduced, with permission, from Del Vecchio et al., Exp. Mycol. *2*, 138–144 (1978)]

centrifuged for 10 min at 10,000 g, and the supernatant was assayed by IEM immediately by the procedure diagrammed in Figure 2. Results using IEM have been published in detail elsewhere (Del Vecchio et al., 1978) and are shown below in Figures 3–6.

3 Counter-Immunoelectrophoresis

In this procedure, glass microscope slides were coated with 1% Noble agar followed by application of one milliliter of 1% agarose (high EEO). Two opposing rows of wells at a distance of 0.5 cm apart were cut out from the agarose. One microliter of cell extract was placed in wells of the row near the cathode (–). The same quantity of antiserum was used to fill wells located on the anodal (+) side. Two microporous wicks were placed lengthwise on each side of the slide to connect the gel to the electrode chamber containing Tris-succinate buffer. CIEP was carried out at 300V for 3 min. The great advantage of the CIEP assay is rapidity, as precipitin lines form within 3 min (Fig. 7).

Electron microscopic visualization of particles in individual precipitin lines was accomplished by cutting out each line, placing these lines in several drops of distilled water, and grinding in an all-glass micro-homogenizer. One drop of this preparation was placed on a Formvar-coated carbon-backed grid for 2 min followed by staining with 1% phosphotungstic acid (PTA) and examination with the electron microscope (Fig. 8).

4 Immunoelectrophoresis

The methodology used for IE has been described previously in detail with reference to antiserum to detect dsRNA in mushrooms (Del Vecchio et al., 1977). These results have been extended, and a series of some ten antisera to VLP of *Agaricus bisporus* has been evaluated.

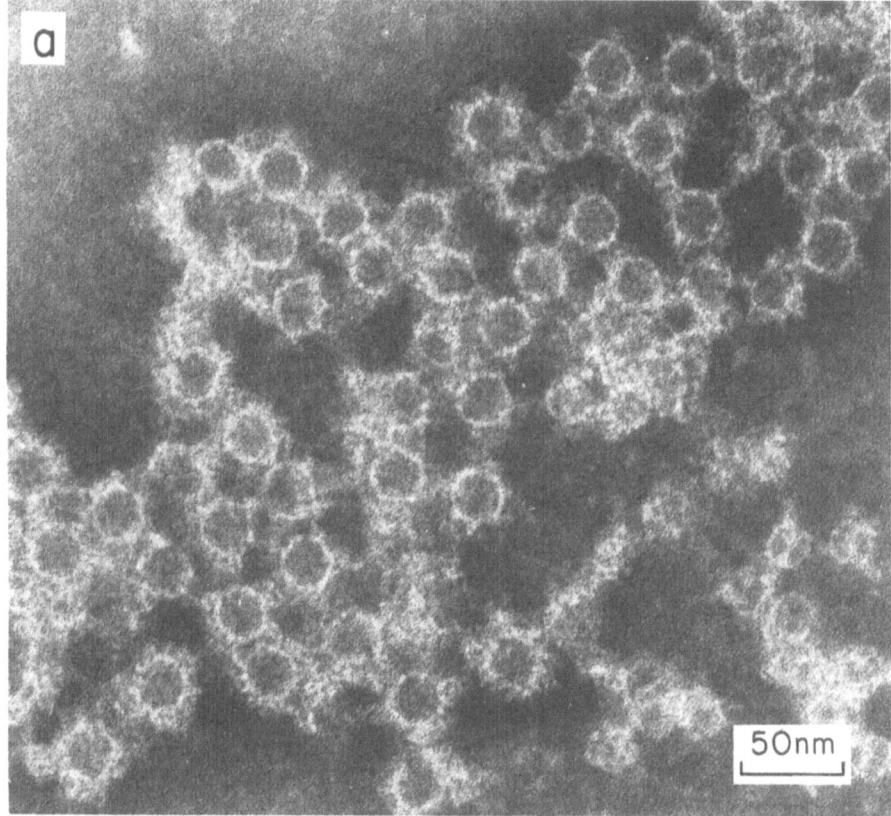

Fig. 4a, b. Immune electron micrographs: a IEM assay of a 100-fold dilution of VLP (25 nm) in an ultracentrifuge fraction. [Reproduced, with permission, from Del Vecchio et al., Exp. Mycol. *2*, 138–144 (1978)]

Samples from ultracentrifuge fractions are electrophoresed on 0.5% agarose film with Tris-sodium barbital buffer (0.06M, pH 8.8) at 60V for 40 min. After electrophoresis, lateral troughs in the agarose are filled with antiserum, and the film is incubated for 24 h at room temperature in a humid environment. The film is washed repeatedly in 0.85% saline to remove excess protein, and the precipitin lines are intensified by placing the film in 95% ethanol for 15 min. Precipitin lines, as in the case of CIEP, can be checked by electron microscopy for particle content (Fig. 9).

5 Enzyme-Linked Immunosorbent Assay

The microplate method of the ELISA followed closely that outlined by Voller et al. (1976a, b) and is diagrammed stepwise in Figure 10. After each step, wells of the microplate are washed 3 times with 0.05% Tween-20. To date, only antisera to 25 nm VLP have been used successfully with ELISA (Fig. 11).

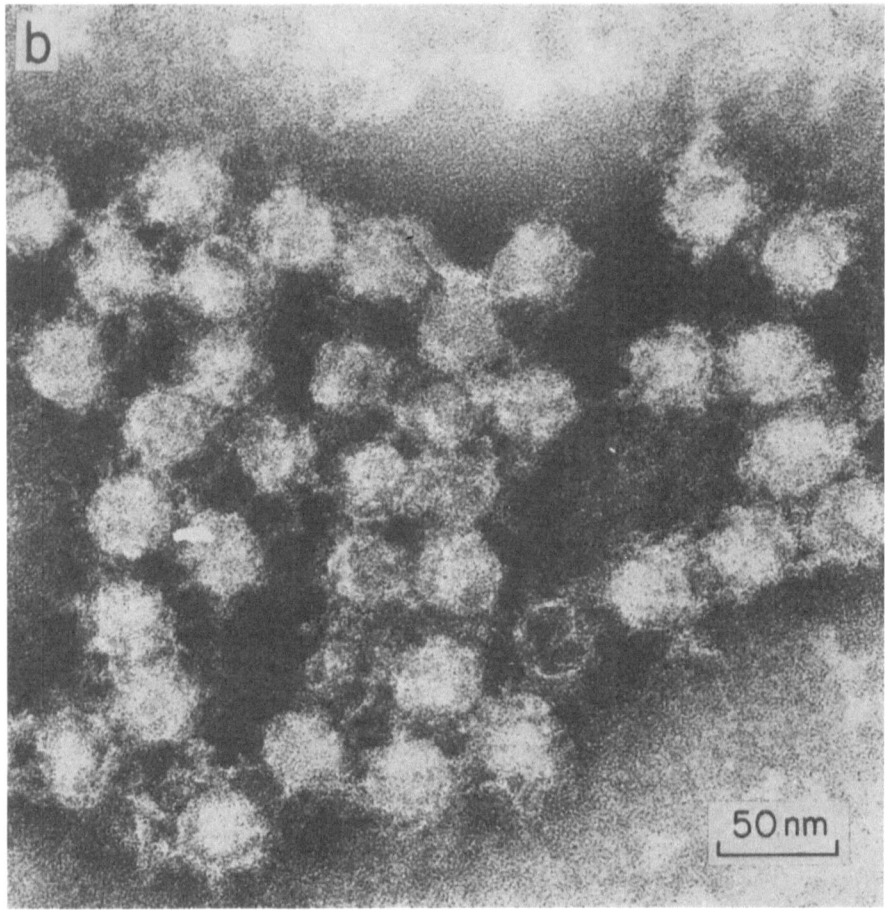

Fig. 4b. IEM assay of VLP (34 nm) that were isolated from a crude extract of an apparently normal mushroom from cultivation

6 Discussion

Four assays for VLP of *Agaricus bisporus* were used in this investigation, immune electron microscopy (IEM), counter-immunoelectrophoresis (CIEP), immunoelectrophoresis (IE), and enzyme-linked immunosorbent assay (ELISA).

The IEM technique resulted in a 50- to 100-fold increase in sensitivity over conventional electron microscopy. Rapid screening of *Agaricus bisporus* extracts was made possible by a specific immunochemical affinity which traps VLP on grids and also allows extensive washing which decreased the amount of cellular debris. The IEM method allowed diagnostic assessment of vegetative cells and mushrooms from normal and diseased strains of *Agaricus bisporus*. An increased number of VLP were detected in

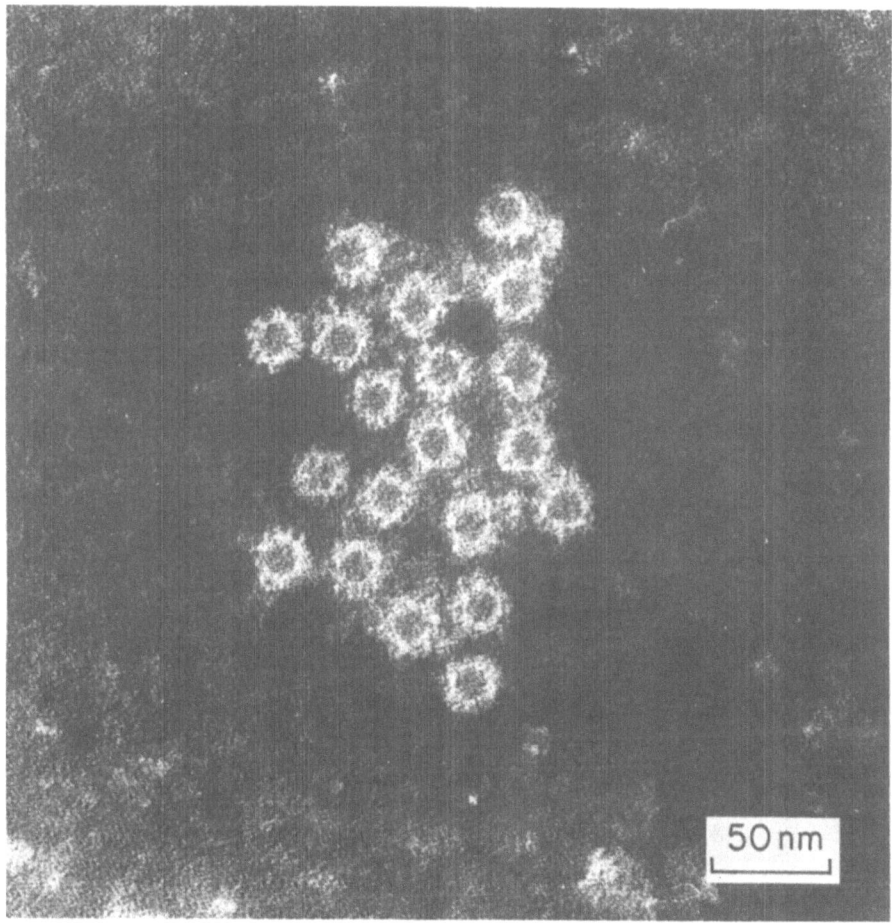

Fig. 5. IEM assay of VLP (25 nm) derived from a crude extract of vegetative cells from a strain of *Agaricus bisporus* grown under sterile laboratory conditions and without disease symptoms in cultivation. [Reproduced, with permission, from Del Vecchio et al., Exp. Mycol. *2*, 138–144 (1978)]

diseased strains, and higher titers were found in mushrooms at the end of the cultivation cycle.

The use of anti-dsRNA in IEM offers a general qualitative test for the presence of viruses which have dsRNA genomes. This technique does presume, however, that dsRNA has its antigenic sites exposed to complex with antibody.

CIEP offers a rapid technique for screening of VLP. Precipitin lines can be seen within 3 min as opposed to 24 to 48 h when other immunoelectrophoretic and immunodiffusion techniques are employed. One microliter samples of VLP extracts and antisera are needed with CIEP, thus rendering this method extremely economical

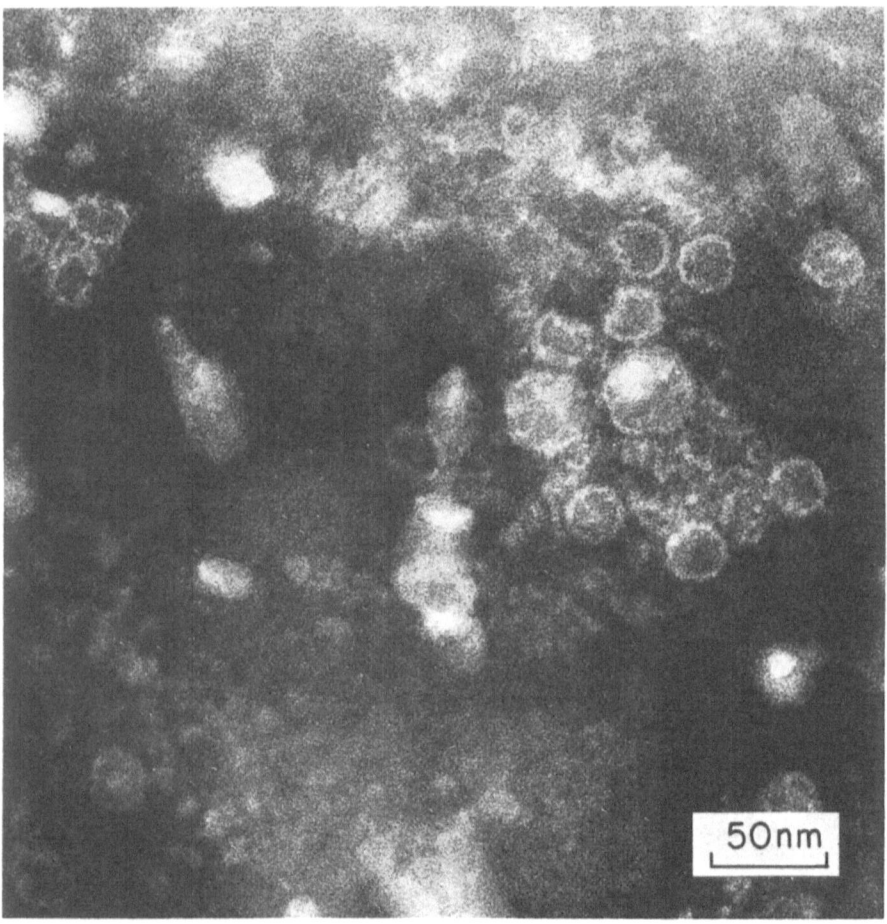

Fig. 6. VLP (25 nm as well as 34 nm) retained on grid prepared with anti-poly I:poly C serum. [Reproduced, with permission, from Del Vecchio et al., Exp. Mycol. *2*, 138–144 (1978)]

in terms of these materials. The rapid nature of this technique also facilitates evaluation of procedures for virus purification, for there is a degree of concentration and specificity for the type of VLP observed, and specific precipitin lines can be defined for VLP content by electron microscopy.

IE can be used to monitor purification of the VLP and has established serological heterogeneity among isometric (25 nm) particles in ultracentrifugation fractions. Like CIEP, IE can be coupled with electron microscopy.

The ELISA is a quantitative assay and the most sensitive assay for VLP to date. It has been used to detect 25 nm particles in diseased but also in symptomless mushrooms.

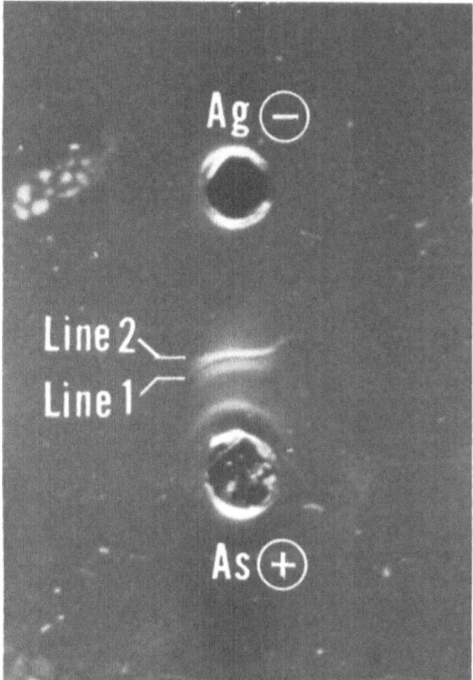

Fig. 7. CIEP of *Agaricus bisporus* VLP antigens *(Ag)* and rabbit anti-VLP antiserum *(As)* showing two virus-specific lines of precipitation

The now confirmed presence of VLP (25 nm and 34 nm) in symptomless mushrooms and in laboratory-grown vegetative cells of *Agaricus bisporus,* while admittedly found therein at low titer, suggests that these viruses in this system, like viruses in most other fungal systems, may be basically latent (Lemke and Nash, 1974). Accordingly, viral disease symptoms and the accumulation of viruses in *Agaricus bisporus* may be a consequence of cultivation, an artifact of the way mushrooms are grown. Mushroom mycelia under cultivation are maintained in compost for extended periods, well beyond the period for maximal rate of vegetative growth. During this time overall synthesis of virus might accumulate relative to cell mass; mushrooms are produced therefrom at an unnaturally high concentration; and mushroom spores carrying virus particles also accumulate to unnatural levels. Such containment of vegetative cells in the presence of a high spore load and the recycling of these cells with higher-than-normal virus titer could lead to further buildup of virus titer and the progressive onset of a disease. As has been recognized previously, the control for mushroom virus disease rests fundamentally upon the practice of good hygiene, the maintenance of spore loads at minimal levels and the adoption of stringent measures to prevent any recycling of *Agaricus bisporus* cells (Dieleman-van Zaayen, 1972).

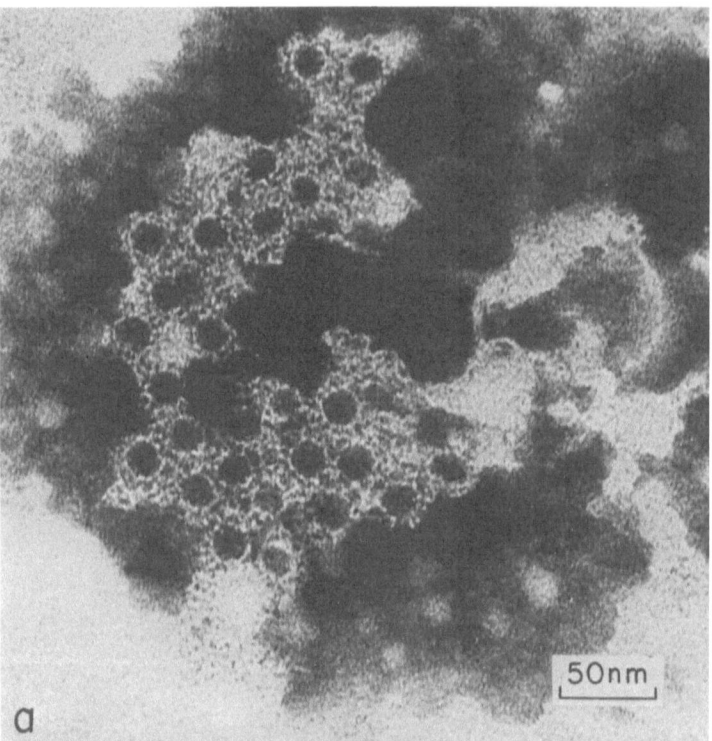

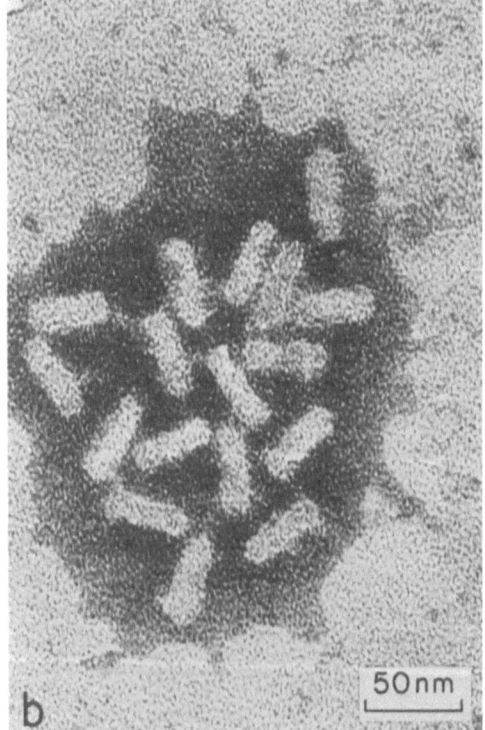

Fig. 8. Electron microscopy of CIEP precipitin lines of Figure 7: a spherical
(25 nm) VLP-antibody complex from
Line 2 of CIEP; b bacilliform VLP-
antibody complex from Line 1 of CIEP

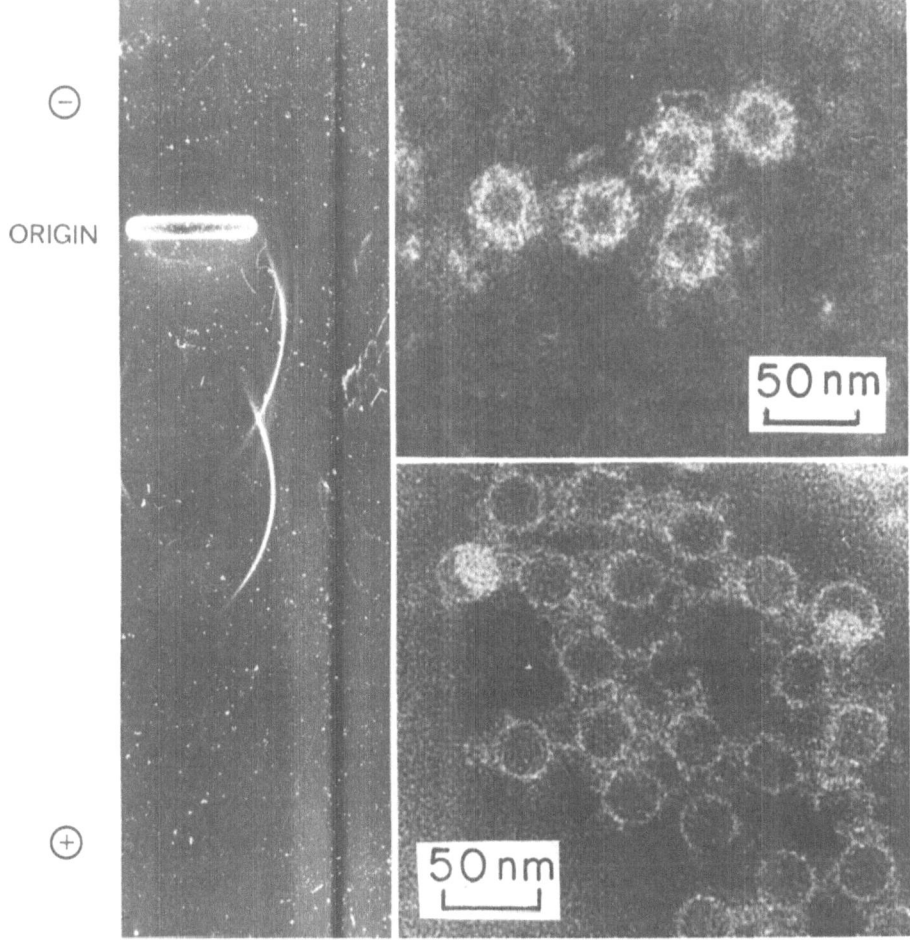

Fig. 9. IE *(left)* and electron microscopy assays *(right)* of an ultracentrifuge fraction containing 25 nm particles. Results indicate two serological types of isometric particles

Summary

The mushroom virus disease of *Agaricus bisporus* is associated with the presence of double-stranded RNA (dsRNA) and several viruslike particles (VLP), two of which having diameters of 25 nm and 34 nm and one being rod-shaped with overall dimensions of 19 X 50 nm. Antisera to these VLP and to dsRNA have been prepared. An investigation of the VLP has been carried out by immuno-electrophoresis, counter-immunoelectrophoresis, immune electron microscopy, and the enzyme-linked immunosorbent assay (ELISA). Immunochemical profiles of the VLP and of dsRNA from various mushroom tissues have been determined in an attempt to relate virus titers to disease expression.

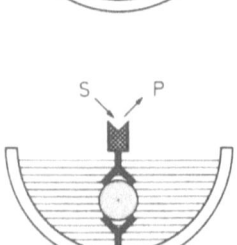

Fig. 10. Procedural steps for ELISA: *(top)* wells are filled with 50 μl of antibodies *(IgG)* in sodium carbonate buffer (0.05 M, pH 9.5) at 37°C for 3 h. IgG nonspecifically binds to polystyrene surface; *(second from top)* serial dilutions of antigens *(VLP)* are added as 50 μl aliquots to wells and microplates are stored overnight at 4°C; *(third from top)* antibodies (IgG) conjugated with enzyme *(Enz)* are added and microplates are allowed to stand for 4 h at 25°C; *(bottom)* enzyme substrate *(S)* is added and the reaction is stopped by addition of 3 M NaOH. Dilution end points for enzyme activity are estimated either visually or spectrophotometrically as intensity of color change for product *(P)*

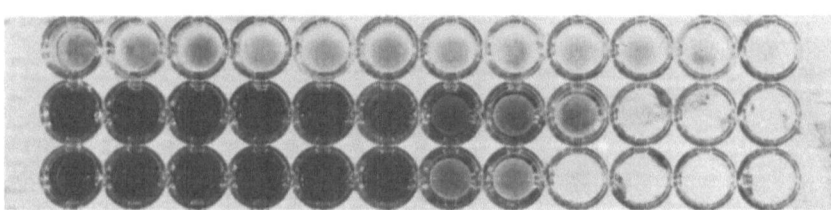

Fig. 11. ELISA microplate employing IgG to 25 nm particles and the enzyme alkaline phosphatase conjugated with particle-specific IgG. The enzyme substrate elicits a yellow product in the reaction. End point of development can be related to 25 nm particle titer *(two bottom rows)*

References

Del Vecchio VG, Dixon C, Lemke PA (1977) Immunoelectrophoretic detection of double-stranded ribonucleic acid from *Agaricus bisporus*. Exp Mycol 1:102–106

Del Vecchio VG, Dixon C, Lemke PA (1978) Immune electron microscopy of virus-like particles of *Agaricus bisporus*. Exp Mycol 2:138–144

Dieleman-van Zaayen (1972) Mushroom virus disease in the Netherlands: symptoms, etiology, electron microscopy, spread and control. Centre for Agric Publ and Document, Wageningen, Netherlands, p 130

Hollings M (1962) Viruses associated with a die-back disease of cultivated mushroom. Nature (London) 169:692–695

Lemke PA (1976) Fungal viruses and agriculture. In: Romberger JA (ed) Virology in agriculture. Allanheld Osmun, Montclair, New York, pp 159–175

Lemke PA, Nash CH (1974) Fungal viruses. Bacteriol Rev 38:29–56

Milne RG, Luisoni E (1977) Rapid immune electron microscopy of virus preparations. Methods Virol 6:265–281

Saksena KN (1975) Isolation and large-scale purification of mushroom viruses. Dev Ind Microbiol 16:134–144 (1975)

Voller A, Bartlett A, Bidwell DE, Clark MF, Adams AN (1976a) The detection of viruses by enzyme-linked immunosorbent assay (ELISA). J Gen Virol 33:165–167

Voller A, Bidwell D, Bartlett A (1976b) Microplate enzyme immunoassays for the immunodiagnosis of virus infections. In: Rose NR, Friedman H (eds) Manual of chemical immunology. Am Soc Microbiol, Washington, DC, pp 506–512

Serological Screening for Fungal Viruses *

R.M. LISTER

Dept. of Botany and Plant Pathology, Purdue University, W. Lafayette, IN 47907/USA

1 Introduction

Infectivity tests cannot readily be used to detect mycoviruses, and the basic method of screening for them is electron microscopy. Indeed, by definition this must clearly be the criterion for establishing the presence of viruslike particles (VLP's) in fungi. However, though electron microscopy must play a key role in any screening program for mycoviruses it presents difficulties, and the use of serological testing can provide valuable supplementary information.

Electron microscopy typically involves tissue sectioning, or the use of empirical methods for concentrating and partially purifying extracts, before detailed examination. Unfortunately, VLP's are often difficult to detect or differentiate against the background of host components, or they may be difficult to isolate in adequate amounts for visualization.

Thus, VLP's such as those of *Penicillium stoloniferum* and *P. chrysogenum*, which reach concentrations of over 1 mg/g dry weight of mycelium (Hollings, 1978) and have a typical isometric viral outline, are quite easy to visualize either in extracts or in tissue sections (Hooper et al., 1972). However, mycoviruses more often occur at about 1–10 µg/g dry weight of mycelium or less (Hollings, 1978), and this low concentration, along with abundant cellular organelles and other constituents, can make visualization difficult. Even VLP's that typically reach adequate concentrations for visualizing in the electron microscope may be difficult to detect in certain situations. For example, *Agaricus bisporus* VLP's MV-1 and MV-2 can regularly be detected in sporophores from chronically infected mycelium, but not in the low concentrations that occur in diseased sporophores from recent infections (Hollings and Stone, 1971).

Again, detecting VLP's simply by screening by electron microscopy presupposes that they will have reasonably conventional sizes and shapes. Although the great majority of VLP's so far reported are isometric particles generally ranging between 25 and 45 nm in diameter (Hollings, 1978), and these characters seem likely to typify mycoviruses, they could in some measure reflect the predisposition of the observer in searching for conventional viral particle types. The various "unusual" particle types, ranging from bacilliform tubes to tailed phage-like particles (see Hollings, 1978), and including the somewhat pleiomorphic club-shaped particles recently associated with hypovirulence in *Endothia* (Dodds, in press; Day and Dodds, 1979) and a disease

---

* Journal paper No. 7376 of the Purdue Agricultural Experiment Station.

in *Agaricus* (Lesemann and Koenig, 1977), may be more widely representative of mycoviruses than is currently suspected.

For several reasons, therefore, serological testing can be useful either as an adjunct to electron microscopy, or as an alternative primary approach to screening cultures for mycoviruses known to be serologically detectable. This contribution is a brief overview of some serological methods that can aid in screening for mycovirus infections in fungi.

2 The Basis of Serological Screening

2.1 Serological Screening Using Antisera to VLP's

Specific antisera have been produced for many VLP's; but because of their specificity they are not widely useful in screening for mycoviruses, most of which appear to be unrelated. However, in establishing relationships (Table 1), or for identifying specific

Table 1. Some serological relationships reported amongst VLP's [a]

Host	VLP's showing relationships
Agaricus bisporus *Agaricus campestris*	25 and 30 nm isometric
Aspergillus foetidus *Aspergillus niger*	Isometric electrophoretically "fast" and "slow" VLP's respectively
Cochliobolus miyabeanus *Helminthosporium sacchari*	30 nm isometric
Diplocarpon rosae *Penicillium stoloniferum*	34 nm isometric and *P. stoloniferum* "slow" virus
Fusarium roseum *Sclerotium cepivorum*	30 and 45 nm, isometric 25 and 40 nm, isometric
Gaeumannomyces sp. nov. *Phialophora radicicola var. graminis*	27 and 35 nm, isometric
Helminthosporium sacchari *Lentinus edodes*	17 x 100–1500 nm, tubular [b]
Penicillium brevi-compactum *Penicillium chrysogenum* *Penicillium cyaneo-fulvum*	35–40 nm, isometric
Penicillium funiculosum *Penicillium purpurogenum*	Isometric

a   Data from Hollings (1978)
b   Particles also reported as protein only (Mori and Kuida, 1977)

types of VLP's associated with specific modifications of biological activity – for example toxin production (Sanderlin and Ghabrial, 1978) – use of antisera to VLP's is clearly indicated. It should be noted however, that even when screening for seemingly similar VLP's in isolates of the same organism, variability may be such that use of one antiserum could fail to detect some VLP types. For example in a comparison of VLP's from 19 isolates of *Gaeumannomyces graminis*, Frick and Lister (1978), found various serotype groupings and degrees of relatedness (Table 2; Fig. 1) indicating that even among VLP's falling into similar size categories, having similar density ranges in CsCl, and isolated from the same organism in the same location, significant serotype differences can occur.

Table 2. Serological screening of *Gaeumannomyces graminis* isolates for dsRNA and titration of three antisera with VLP's from positive isolates. (Data from Frick and Lister, 1978)

Isolate	Source [a]	dsRNA	Antiserum titer [b]		
			No. 3	No. 59	M-10
No. 59	C.J. Rawlinson-Oats, England	+	64	1024	0
ATCC 12761	American Type Culture Collection				
	Rockville, Maryland	+	0	0	2
No. 3	Wheat-1973 Vincennes, Indiana	+	4096	256	0
M-10	Wheat-1975 Vincennes, Indiana	+	0	0	512
G-1	Wheat-1975 Vincennes, Indiana	+	4096	128	4
E-1	Wheat-1975 Vincennes, Indiana	+	512	No data	No data
E-3	Wheat-1975 Vincennes, Indiana	+	512	256	0
F-2	Wheat-1975 Vincennes, Indiana	+	64	0	8
B-3	Wheat-1975 Vincennes, Indiana	+	0	0	8
S-6	Wheat-1975 Vincennes, Indiana	+	0	0	64
M-5	Wheat-1975 Vincennes, Indiana	+	0	0	128
M-13	Wheat-1975 Vincennes, Indiana	+	0	0	512

a   All culture isolates from Vincennes, Indiana were collected from the same field which had been cropped to wheat for 11 years

b   Antiserum titers are presented as reciprocals and were determined with unfractionated, concentrated VLP preparations suspended in 1 ml 0.01 M Tris-HCl (pH 8.5)/5 g of starting mycelium. Antisera were made to VLP's from isolates No. 3, No. 59, and M-10 collected from similar VLP-containing fractions of CsCl gradients

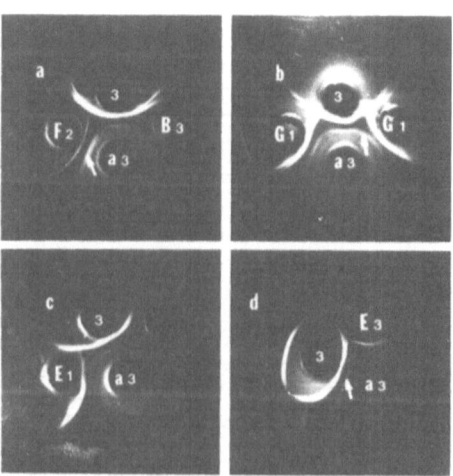

Fig. 1 a–d. Agar gel diffusion serological comparisons made using antiserum to VLP's from *Gaeumannomyces graminis* isolate No. 3 (a3) with unfractionated VLP preparations from isolates: **a** *F-2* and *B-3;* **b** *G-1;* **c** *E-1;* **d** *E-3* and similar preparations from isolate 3 (for details of isolates see Table 2). Note spurring reactions between various isolates and No. 3 VLP's when placed in adjacent wells, confirming different degrees of serological relatedness. Nonspecific reactions (indicated by *arrows* in **a, b,** and **d**) occurred close to antiserum wells only when concentrated antiserum was used

Because many mycoviruses present special difficulties in purification from contaminating antigenic host proteins and polysaccharides, there is also always a risk that use of antisera to VLP's will give spurious positive results or indications of relationship. VLP antisera frequently contain antibodies to normal host constituents (Fig. 1) and cross-absorption with extracts from VLP-free cultures may be a prerequisite for their use (Moyer and Smith, 1976, 1977).

Caution in assessing relationships suggested by tests with VLP antisera is also required for another reason: for two dsRNA plant viruses (rice dwarf and maize rough dwarf viruses) it has been shown that antisera elicited by injecting rabbits with viral particles can contain antibodies to dsRNA as well as capsid protein (Ikegami and Francki, 1973). Cross-reactions were shown to result not from precipitation of viral proteins but of dsRNA. This effect does not always seem to have been excluded in examining relationships indicated by VLP antisera. However, it may be unusual, for it was not found in the studies of *G. graminis* relationships mentioned above. Antisera to VLP's from *G. graminis* and *P. stoloniferum* did not cross-react in reciprocal tests, nor did several antisera to fungal VLP's react with poly I:poly C or polyA:poly U preparations (Frick and Lister, 1978). Presumably in this situation the elicitation of antibodies recognizing dsRNA depends on the stability of the viral capsids during antibody production, and their ability to promote hapten activity of the dsRNA.

2.2 Serological Screening Using Antisera to dsRNA

DNA- and ssRNA-containing fungal VLP's have been reported, but most mycoviruses so far examined contain segments of dsRNA (see Hollings, 1978). Techniques for producing antisera specific for RNA double-strandedness, by injecting rabbits with synthetic polyribonucleotides, have been available for some time (e.g., Lacour et al., 1968; Schwartz and Stollar, 1969). In addition to reacting with the synthetic dsRNA's, such antisera have been shown to detect dsRNA's in viruses of animals (Schwartz and Stollar, 1969) and plants (Francki and Jackson, 1972). In my laboratory, such antisera to synthetic dsRNA's detected mycovirus infections (Moffitt and Lister, 1973, 1975; Frick and Lister, 1978) and our antisera have also successfully been used by others for mycovirus detection and the specific detection of dsRNA (e.g., del Vecchio et al., 1977; Sanderlin and Ghabrial, 1978; Derrick, 1978).

2.2.1 Preparation of Antisera to dsRNA

The procedure for developing antisera that react with dsRNA is quite simple, and derives from the introduction of methylated bovine serum albumin as a complexing agent allowing dsRNA to act as a hapten (Plescia et al., 1968). The following quotation from Moffitt (1973) may be useful to those unfamiliar with the process:

"Antisera were made to complexes of synthetic dsRNA and methylated bovine serum albumin (MBSA) as described by Francki and Jackson (1972). Bovine serum albumin, fraction V (BSA) (Miles Laboratories) was methylated by the method of Ralph and Bergquist (1967): 5 g BSA were suspended in 500 ml A.R. absolute methanol and 4.2 ml 12 *N* HCl were added. The precipitate which formed when the solution was incubated at 37°C for 3 days was collected by centrifugation, washed twice with methanol, and twice with anhydrous ether. The final dry product was dissolved at a concentration of 2 mg/ml in distilled water, and the solution was then adjusted to 1 mg/ml in SSC (0.15 M NaCl, 0.015 M Na citrate pH 7.2) by the addition of an equal volume of SSC at twice the standard concentration.

Synthetic double-stranded polyribonucleotide, either poly I:poly C or poly A:poly U (Miles Laboratories) at 1 mg/ml in SSC, was complexed with MBSA at 1 mg/ml by mixing equal volumes of the two solutions at room temperature. The resulting insoluble floc was emulsified with an equal volume of Freund's incomplete adjuvant (Difco). Three intramuscular injections containing 800, 1000, and 10,000 μg nucleotide respectively were given at weekly intervals to New Zealand white rabbits. In the fourth week, blood samples were collected from the ear vein, and then 1000 μg polynucleotide complexed with MBSA was injected intravenously. Blood samples were collected at intervals thereafter, and intravenous "booster" injections were also given."

Table 3. Reciprocal titers of antisera to dsRNA as tested against homologous and heterologous antigens in immunodiffusion tests [a]

Time after last primary injection			Antiserum			
			poly A:poly U		poly I:poly C	
	Weeks	Bleed No.	Antigen tested [b]		Antigen tested [b]	
			poly A: poly U	poly I: poly C	poly A: poly U	poly I: poly C
	0	1	8 (16)	4 (2)	4 (2)	4 (4)
	0.5	2	8 (32)	8 (8)	2 (4)	8 (8)
	1	3	8 (32)	4 (8)	4 (16)	8 (8)
Booster	2	4	8 (32)	8 (2)	4 (16)	8 (8)
injection	4	5	16 (16)	8 (2)	4 (8)	4 (4)
	6	6	8 (16)	4 (0)	4 (8)	8 (4)
	11	7	8 (16)	2 (0)	2 (4)	16 (2)
Booster	16	8	8 (8)	2 (0)	16 (2)	16 (4)
injection	26	9	16 (2)	2 (0)	32 (0)	32 (0)

a   Data from Moffitt (1973) and Frick (1976)
b   Antigens were tested at 1 mg/ml (Moffitt) or 0.1 mg/ml (Frick-data in parentheses)

Table 3 gives the titers of test bleeds made at intervals after injections in two separate sets of experiments (Moffitt, 1973; Frick, 1976), as determined in two-dimensional double-diffusion tests in an agar. All titers were relatively low, although titers determined in ring precipitin tests were higher: for example for Moffitt's bleed 5 (antipoly A: poly U) in Table 3, they were 1/128 and 1/256 against poly A:poly U and poly I: poly C respectively, and for bleed 3 (antipoly I:poly C) they were 1/128 and 1/128 respectively. The minimum detectable amounts of synthetic polynucleotides in gel diffusion tests for both types of antiserum were about 10 μg/ml of poly A:poly U and 4 μg/ml of poly I:poly C. As little as 0.6 μg/ml of the polynucleotides were detectable in ring precipitin tests. Titers may be improved, for example by use of the complete adjuvant, and the useful titer of an antiserum depends of course on the serological test procedure selected.

2.2.2 Specificity

dsRNA antisera produced as above have typically reacted reciprocally to both artificial polynucleotide duplexes, but homologous titers are often higher than heterologous titers (Table 3). Moreover, homologous precipitin lines are frequently more sharply defined than those of heterologous reactions, which frequently consist of several precipitin lines (Fig. 2). When antipoly I:poly C reacted with polyI:poly C and poly A: poly U in adjacent wells, homologous and heterologous precipitin lines fused without

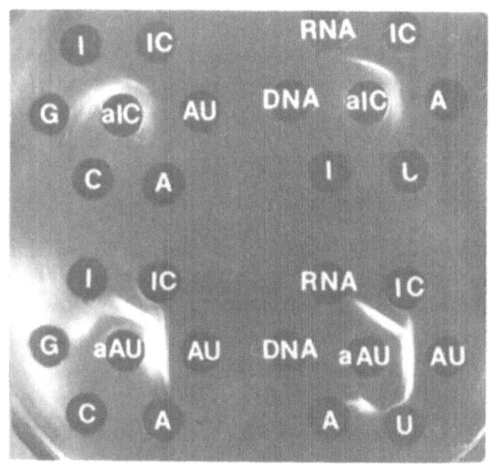

Fig. 2. Gel diffusion tests showing specificity of antisera made against dsRNA. Antisera made against either poly I: poly C (aIC) or poly A:poly U (aAU) are in center wells. Antigens tested: poly I (I), poly I:poly C (IC), poly A;poly U (AU), poly A (A), poly C (C), poly G (G), poly U (U), calf thymus DNA and E. coli ribosomal RNA, all at 1 mg/ml

spur formation; but in the reciprocal situation, spur formation occurred indicating the presence of antibodies on the antipoly A:poly U antiserum that were not present in the antipoly I:poly C antiserum (Fig. 2). Neither kind of antiserum reacted with the single-stranded polynucleotides poly A, poly U, poly I, poly C, or poly G, or with *Escherichia coli* RNA, calf thymus DNA, or RNA from TMV (Fig. 2; Moffitt and Lister, 1975), and this list was later extended (Frick and Lister, 1978) to include poly X, poly A:U, poly U:C, poly A:I, poly U:I, poly A:U:C, *Blastocladiella* RNA, yeast ssRNA, salmon sperm DNA, and herring sperm DNA. However, where ssRNA preparations appropriate for annealing were placed in adjacent wells in gel diffusion tests, precipitin reactions often occurred (Fig. 2; Moffitt and Lister, 1975; Frick, 1976), an effect also noted by Schwartz and Stollar (1969). Thus, all the results agreed in suggesting that the antisera were highly specific for both double-strandedness and the ribose sugar moiety, while also suggesting caution in the arrangement of samples in gel-diffusion tests.

    In contrast to this, cross-reactivity has often been reported between dsRNA and antisera and mononucleotides or DNA (Table 4). Leaving aside the question of antigen

Table 4. Specificity of antisera to polyribonucleotides as determined in gel diffusion tests. (Data from Moffitt, 1973)

Antigen tested	Immunizing Antigen					
	poly A:poly U		poly I:poly C			
	Schwartz and Stollar, 1969	Lacour et al., 1968	Lacour et al., 1968	Schur and Monroe, 1969	Francki and Jackson 1972	Field et al., 1972
poly A:poly U	+	+	+	+		+
poly I:poly C	+	+	+	+	+	+
poly G:poly C	−	−	−	−	−	
poly A	−	+	−	−		−
poly U	−	−		−		−
poly I	−	−	+	+		0
poly C	−	−	−	−		−
poly G	−	−	−	−		−
RNA	− a	+ c	+ c	− e	− g	− i
DNA	− b	+ d	−	− f	− h	− j

a   Escherichia coli soluble RNA, yeast RNA, total RNA from KB cells
b   Native or denatured DNA from calf thymus, salmon, E. coli, dogfish erythrocytes
c   Native or denatured total, ribosomal and transfer RNA from mouse ascites cells. Not all antisera which were tested reacted
d   Native or denatured DNA from Micrococcus lysodeikticus, phage 2 C, calf thymus, Clostridium perfringens. Not all antisera which were tested reacted
e   F2 phage RNA, poliovirus RNA, yeast RNA, E. coli soluble RNA, rat liver soluble RNA, rat and rabbit liver ribosomal RNA
f   Native or denatured DNA from calf thymus
g   Tobacco mosaic virus RNA, tobacco leaf RNA, yeast RNA, sugar cane leaf RNA
h   Tobacco leaf DNA
i   Yeast RNA
j   Calf thymus DNA

purity, there is really no conflict in these results. As reviewed by Lacour et al. (1968), several levels of specificity are possible in serological reactions involving polynucleotides, including sugar type (ribose or deoxyribose), base type (purine or pyrimidine), macromolecular structure (strandedness, nucleotide arrangement, left- or right-handed helices), and the precise determinants exposed by the binding arrangement of the polynucleotide with its carrier MBSA.

Complexity in specificity is suggested by the spurring reaction noted with anti-poly A:poly U serum (Fig. 2) and in the occurrence of multiple precipitin lines frequently observed in gel diffusion tests (Fig. 2 and see Lacour et al., 1968), although differences in the diffusability of the dsRNA segments from different VLP's might also cause this effect. Differences between antisera due to subtle differences in methodology are also possible. Again, antisera made with artificial dsRNA's from different sources could differ, and certainly this would be expected with antisera made with natural dsRNA's such as dsRNA prepared from mycoviruses. This effect is already indicated by work of R.I. Barton (reported in Hollings, 1978) in which the dsRNA from P. chrysogenum reacted much more strongly with its homologous antiserum than with antiserum to poly I:poly C.

Though Plescia et al. (1969) suggest that quite small regions of polynucleotide duplexes – about 5 nucleotide pairs long – may be adequate to form an antibody-binding site, the behavior of antisera used in screening fungi for dsRNA's will depend on the characteristics of their predominating antibodies. Detectability of cross-reactions will also depend on the sensitivity of the test procedures used (e.g., gel diffusion versus complement fixation).

2.3 Extraction of dsRNA from VLP's and Fungi

2.3.1 Extraction of RNA from VLP's

Many methods are available for extracting nucleic acids from viral particles. With us, the single-phase phenol-ethanol methods of Diener and Schneider (1968) have proved quite successful (Moffitt and Lister, 1975; Frick and Lister, 1978), as have slight modifications of the two-phase system, involving phenol-SDS extractions at room temperature followed by ethanol precipitation (see Ralph and Berquist, 1967).

2.3.2 Extraction of RNA from Mycelium: Approach 1

For extracting total nucleic acid including dsRNA, from mycelium, we have used a modification of the procedure used by Francki and Jackson (1972). Mycelium is harvested by filtration from culture solution and washed. A sample of the mycelial cake (usually 10 g) is thoroughly extracted by blending in a glass Waring blender, at room temperature, with a mixture of water-saturated phenol containing 0.1% hydroxyquinoline (4 volumes on a w/v basis), TNE buffer (0.1 M Tris-HCl, 0.1 M NaCl, 0.01 M EDTA, pH 7.0; 3 volumes), 4% SDS (1 volume) and EDTA-washed bentonite at 40 mg/ml (0.4 volumes). Yields are improved by reextracting the phenolic and denatured phases, separated from the aqueous phase by low speed centrifugation (20 min at 8,000 g), with a further TNE buffer-phenol mixture. The combined aqueous phases are reblended with equal volumes of water-saturated phenol and a chloroform:butanol mixture (1:1, v/v). Nucleic acids are separated from this clarified aqueous phase by two successive precipitations with 3 volumes of absolute ethanol at – 12°C overnight. The final preparations are suspended in STM buffer (= 0.1 M NaCl, 0.1 M Tris/HCl, 0.01 M MgCl$_2$, pH 7.3) at 1 ml/5–10 g starting weight of mycelial cake.

Using this procedure, and resuspending the nucleic acids at 1 ml/10 g, yielded preparations from VLP-infected G. graminis cultures that reacted up to dilutions of 1/32 against anti poly I:poly C antiserum. Comparison with the titers of poly I:poly C standards indicated a yield of about 0.8 μg/g of dsRNA from the mycelial cake (Frick and Lister, unpublished). With VLP-containing P. chrysogenum cultures, similar preparations reacted to dilutions of 1/128, and yields were approximately doubled by passing the phenol homogenate through a French press, but the shearing forces involved probably fragmented the dsRNA (Moffitt and Lister, 1975).

2.3.3 Extraction of dsRNA from Mycelium: Approach 2

Morris (1977) has developed a batch procedure for extracting and separating dsRNA from virus-infected plant and fungal tissues, based on its differential affinity for cellulose (Whatman, CF-11 or Biorad N-1) in aqueous ethanol (Franklin, 1966) at a concentration of 15% (Jackson et al., 1971). Sufficient concentration was achieved to detect dsRNA replicative form in some ssRNA plant virus infections (Morris, 1977). We have applied this procedure (Morris, personal communication), slightly modified, to extracting dsRNA from *G. graminis*. Ten gram samples of washed mycelial cake were extracted by blending in a mixture of 20 ml GPA buffer (0.2 *M* glycine, 0.1 *M* Na$_2$HPO$_4$, 0.6 *M* NaCl, pH 9.5), 2 ml 10% SDS, 0.2 ml mercaptoethanol, 20 ml phenol, 20 ml chloroform: pentanol (25:1), at room temperature. Further processing is as above, but dsRNA is specifically separated out from the aqueous phase by making it 15% with ethanol and adding chromatographic cellulose powder at 0.25 g per 20 ml, with gentle stirring for 10 min, on ice. The cellulose, pelletted by low speed centrifugation, is then resuspended in a mixture of 15% ethanol: 85% STE buffer (0.1 *M* NaCl, 0.05 *M* Tris/HCl, 0.001 *M* EDTA, pH 7.0) and thoroughly washed in a cellulose column with further ethanol-buffer mixture, before elution of the dsRNA in STE buffer and precipitation with absolute ethanol at −20°C.

In comparison with "approach 1" described above, this method yielded somewhat less dsRNA from *G. graminis* mycelium, but the nucleic acid was a much purer product, and was resolved better in poly acrylamide gel electrophoresis (Frick and Lister, unpublished).

3 Serological Test Procedures

Similar test procedures have been applied for screening for both VLP's and dsRNA from mycovirus-infected fungi. Methods used are reviewed by Adler and del Vecchio (1979), but some comments are appropriate here. It should perhaps be emphasized that most of these test procedures have been applied only to a few fungal-virus systems. Each system presents its own peculiar problems, and an assay useful for detecting mycovirus infection in one system may not be applicable to another.

3.1 Two-Dimensional Double Diffusion in Gels (Ouchterlony Gel Diffusion Test)

Gel diffusion testing has the advantages of reasonable economy in use of antigen and antiserum, and of differentiating between reacting systems by their diffusion rates. Figure 1 illustrates how this may distinguish between precipitin lines due to VLP's and those due to nonviral host constituents in mycovirus preparations. Spurring reactions (Fig. 1) can also aid in discriminating between isolates. Results can be related to the presence of specific VLP types visualized in the electron microscope.

Similarly, reactions with dsRNA antisera require to be related with electron microscopy. In the initial work with antisera to poly I:poly C and poly A:poly U (Moffitt and Lister, 1975), 20 fungal isolates indexed positive for dsRNA of the 70 selected for testing (Fig. 3). Of these 20, 18 were examined for VLP content by electron microscopy, but VLP's were found in only 5. Reasons suggested for the absence of VLP's in the remainder included the possibility of loss of VLP's in extraction and clarification, low VLP content, viewing difficulties, and the possible existence of unencapsulated dsRNA's. The occurrence of unencapsulated dsRNA's in fungi now seems

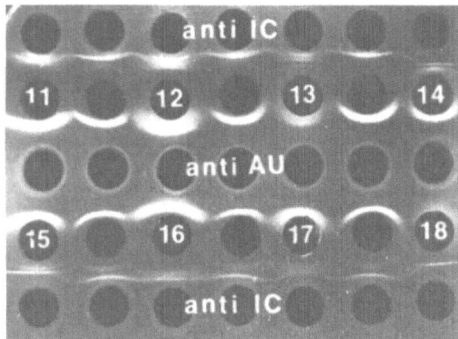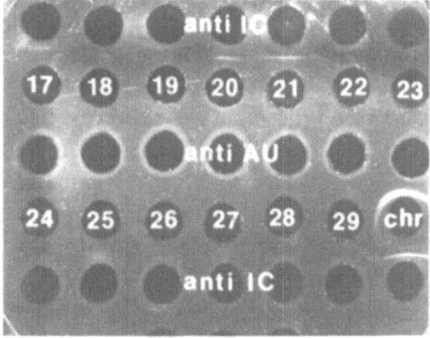

Fig. 3. Selected serological tests indicating the presence or absence of dsRNA in fungal isolates screened as "unknowns". (For complete details, see Moffitt and Lister, 1975). Antisera are in rows marked *antiIC* and *antiAU:* nucleic acid preparations are in numbered wells. The 11 through 18 series on *left* are selected positive reactors. The 17 through 29 series on *right* are mostly selected negative reactors. *Penicillium chrysogenum* VLP dsRNA is in alternate (*unnumbered*) wells as a marker in series on left, and in well marked *chr* in series on right. Unknowns were: *left: 11, H. maydis* race 0, slow-growing; *12, Pythium butleri; 13, Colletotrichum graminicolum*, Miss.; *14, Endothia parasitica* 1503; *15, E. parasitica* 1505; *16, Gaeumannomyces graminis* 3; *17, Penicillium funiculosum* RC 2; *18, P. cyclopium* RC 15 (all indexed positive with one or both antisera); *right: 17, Phyllosticta; 18, Kabatiella zeae; 19, Colletotrichum coccoides; 20, Fusarium roseum; 21, Colletotrichum falcatum; 22, H. carbonum* race II, 51; *23, H. carbonum* race II, 20; *24, Gaeumannomyces graminis* 2; *25, Endothia parasitica* 1502; *26, E. parasitica* 1504; *27, Aspergillus flavus* 10124; *28, A. flavus* 18166; *29, A. flavus* 15546. Note that isolates *20* and *21* reacted with antiIC only

well established (e.g., Koltin and Day, 1976). Of special interest, in view of recent work, was that the two hypovirulent *Endothia* isolates tested (Nos. 14 and 15 in Fig. 3 left) were strongly positive for dsRNA, whereas the two virulent isolates were not (Nos. 25 and 26 in Fig. 3 right), although VLP's were not found in any isolate. The occurrence of dsRNA in hypovirulent *Endothia* isolates has since been reported by Day et al. (1977), and Dodds (in press, and personal communication) has observed specific club-shaped particles in such isolates.

In the tests illustrated in Figure 3, reactions with antipoly I:poly C were more sharply defined than reactions with antipoly A:poly U, and in a few cases reactions with antipoly A:poly U were negative or doubtful when reactions with antipoly I: poly C appeared positive (e.g., Nos. 20, 21, Fig. 3). These results may reflect differences in optimal proportion ratios of the antigens and antibodies used, or the kind of differences between dsRNA antisera specificities referred to above. Variability in reactions using dsRNA antisera was also encountered in screening *G. graminis* isolates for dsRNA (Frick and Lister, 1978) but the various antipoly I:poly C samples used did not always give clearer reactions than antipoly A:poly U. Probably it is desirable in all such screening work to use several dsRNA antisera.

3.2 Fluorescent Antibody Test

The indirect fluorescent antibody test has been used successfully to detect VLP's and dsRNA in fungal mycelium, spores and yeast cells (Adler, 1974; Adler et al., 1975). Standard procedures are used for the various steps involved in treating samples, first with specific antibodies, and then in detecting the fixed antibodies with fluorescently labeled goat antirabbit gamma globulin. Adler (personal communication) suggests young hyphal growth with minimum sporulation, grown on cellulose strips, as the best type of mycelium for this purpose.

Unfortunately, there are problems in the routine application of this test. Auto-fluorescence occurs in some fungi and in yeast cells (Rose and Adler, 1977), and non-specific binding of the fluorescein-labeled antibody can also occur. In tests of *G. graminis* cultures by this procedure, using antisera to dsRNA (Frick and Lister, unpublished), we have had some success, but the results have been too erratic to rely on unless confirmed by other tests. Our impression is that either the test is relatively insensitive, or that penetration of the hyphae by antibody can be unreliable.

In principle, the test has great potential both as a screening method and for locating virus and viral products within hyphae. At present it seems that the precise methodology required for success may have to be worked out in relation to each fungal type.

3.3 Indirect Agglutination

Adler and co-workers (Ohlson et al., 1977; Rose and Adler, 1977) have developed a sensitive indirect agglutination test for dsRNA, making use of polyacrylamide micro-beads to which antibodies specific for dsRNA can be bound by the dual reaction compound 1-ethyl-3-(3-dimethyl aminopropyl) carbodiimide HCl (EDAC). Clumping of the sensitized beads indicating a positive reaction can occur within 60 s with 15 $\mu g/ml$ of poly I:poly C.

A drawback with this test is the reactivity of the beads with DNA (which may be a reflection of the test sensitivity). Also, the sensitized beads progressively lose reactivity when stored for more than 2 or 3 weeks at 4°C.

It would be of interest to investigate whether other methods of enhancing sensitivity and visualization of serological reactions by using sensitized particles — e.g., latex beads, bentonite, red blood cells (see Ball, 1974) — would be of value in screening for mycoviruses or dsRNA.

3.4 Immunoelectrophoresis

Del Veccio et al. (1977) have successfully applied immunoelectrophoretic analysis to the dsRNA's in mycovirus-infected *Agaricus bisporus,* and Adler (personal communication) has adapted the technique of rocket immunoelectrophoresis as a quantitative assay for dsRNA in fungi. These procedures provide powerful analytical tools for investigating mycoviruses, many of which are complex, multicomponent systems. Especially with the rocket immunoelectrophoresis technique, there is potential for screening many samples simultaneously under the same conditions at the same time in one large gel plate.

3.5 Serologically Specific Electron Microscopy

Derrick (1978) has applied the technique of serologically specific electron microscopy (Derrick and Brlansky, 1976) to the detection of dsRNA (replicative form) in ssRNA plant virus infections. Electron microscope grids were treated with sera specific for dsRNA, and buffer extracts from tobacco mosaic virus-infected plants were then applied and unbound materials removed by washing. Double-stranded RNA was readily detected in infected extracts and also in a solution of artificial poly I:poly C. Using this technique it is therefore to be expected that dsRNA in fungal extracts could be detected similarly. One advantage of the technique, in cases where the dsRNA is not encapsulated is the avoidance of the use of phenol or detergent, because with use of these reagents, complementary ssRNA strands tend to become double-stranded. Derrick suggests also that positive visualization of the dsRNA as extended threads of high molecular weight material obviates the possibility of false positive reactions that might occur in using other serological methods of detecting dsRNA due to short regions of base-pairing in otherwise single-stranded structures.

3.6 Enzyme-Linked Immunosorbent Assay (ELISA)

Since their recent introduction (Voller et al., 1974) ELISA applications have become so commonplace that it seems reasonable to suppose that they will be applicable to most systems for which specific antisera are available. ELISA is based on the enhancement of visualization of serological reactions up to about 1000-fold or more compared with usual precipitin tests, by the use of enzyme-labeled antibody. Precipitin reactions are detected by testing for the presence of bound enzyme by a sensitive enzyme/substrate reaction (usually alkaline phosphatase hydrolysis of p-nitrophenyl phosphate) in tests performed in commercially available polystyrene plates or tubes, which act as an inert carrier for the reaction. The list of routine applications is extensive and growing, and it seems highly likely that application to screening for mycoviruses could become one. However, the results of our recent tests of the ELISA procedure for detecting dsRNA (Frick and Lister, unpublished) suggest that it may be less useful than anticipated for this purpose. Gamma globulins from antipoly A:poly U and antipoly I:poly C did not readily detect the heterologous artificial copolymers at the gamma globulin dilutions required to reduce "background" nonspecific reactions to acceptable levels. As mentioned earlier, each application of a serological test procedure may present special problems and it is too early to decide how useful ELISA may become in this regard. Certainly, improved sensitivity would appear to be an advantage in serological screening for VLP's, but possibly the need to use gamma globulins at high dilutions to reduce "background" reactions, in combination with the specificity problems already alluded to, could be a drawback to applying ELISA to detecting dsRNA's in fungi.

4 General Remarks

At the present stage of development of serological screening methods for mycoviruses
it would be premature to make any definite recommendations. For the present it
appears that, though highly sensitive procedures such as ELISA may be useful in
screening for specific or related VLP's with antisera to highly purified VLP preparations, the use of antisera detecting dsRNA in the various gel diffusion test procedures
may offer the best option for wideranging serological screening for fungal viruses. The
relative insensitivity of these methods is no drawback since the antisera may need to
be used concentrated to detect a wide range of dsRNA species.

Differentiation of dsRNA species and contaminants by differential diffusion and
electrophoretic mobility provides powerful analytical and separatory potential in these
types of test. The possibility of concentrating dsRNA from crude mycelial extracts by
its differential affinity for cellulose in aqueous ethanol makes it realistic to detect
dsRNA in fungi even if present in very small amounts. Indeed, following the results
with plant viruses (Morris, 1977), it may be possible to detect even ssRNA fungal virus
infections through the detection of dsRNA replicative form.

Acknowledgments. I wish to thank Dr. J.P. Adler for allowing me to see the manuscript of her section of a chapter on *Specialized Assay for the Detection of Mycoviruses* to be included in *Viruses
and Plasmids in Fungi*, Lemke, P.A. (ed.) (Marcel Dekker Inc.), also L.I. Carter (née Frick) for
valuable discussions.

Summary

Lacking routinely applicable infectivity tests the techniques generally used for screening fungi for
viruslike particles (VLP's) are electron microscopy (E.M.) and serological testing. E.M. − by definition the basic criterion of VLP detection − typically involves tissue sectioning or empirical methods
for concentrating and partially purifying extracts, before detailed examination. Unfortunately,
VLP's can be difficult to differentiate against the background of host components, or to isolate
in amounts adequate for visualization. Also, detection by E.M. presupposes the VLP will have
reasonably conventional size and shape. Serological testing can thus be a useful adjunct to E.M.
or an alternative primary approach to screening.

The VLP-specific antisera available are not widely useful in screening, because serological
relatedness of distinct mycoviruses is known in only a few instances. Moreover, recent work suggests that even when screening for related VLP's, serotype variability may cause problems. However, most mycoviruses seem typically to contain dsRNA, and antisera that specifically recognize
RNA double strandedness have proved to be rather widely applicable to detecting VLP's in fungi.
Thus, in our serological screening of 70 fungus isolates covering a wide range of genera, 20 reacted
positively with antisera to synthetic ds-ribonucleotides. Of these 20, only 5 were confirmed by
E.M., and it was suggested that this might mean that serological testing could be more sensitive
than E.M., because it bypasses VLP extraction and viewing difficulties. Interestingly, recent work
with *Endothia parasitica* cultures strongly supports this suggestion, for the two hypovirulent cultures that reacted positively for dsRNS in our tests have now been proven by other workers to
contain the "nontypical" VLP's they associate with hypovirulence.

For screening purposes, constraints on the use of the dsRNA screening test as originally set out
include the need to propagate fairly substantial amounts of mycelium (about 10 g), for processing
by phenol extraction and nucleic acid concentration. However, it has been shown that use of
fluorescently labeled antibodies to dsRNA can enable the direct application of serological testing

to fungal hyphae. Another simplification might be the use of the batch and column processes for dsRNA separation and concentration based on cellulose chromatography, which have recently enabled detection of ds (replicative form) RNA's in even ssRNA plant virus infections. Where there is a need to screen for specific types of VLP's – as for example in investigating whether infection with specific VLP's is associated with modified biological activity – antisera to the specific VLP's involved must be used. Here again, direct treatment of hyphae with fluorescently labeled antisera could aid in reducing the need to attempt extraction and concentration of VLP's, but the VLP content of hyphae may well be limiting. Possibly the use of the extremely sensitive enzyme-linked immunosorbant assay (ELISA) procedure currently proving so valuable in plant virology would be appropriate here for testing mycelial extracts.

References

Adler JP (1974) Viruses of yeast. Dev Ind Microbiol 16: 152–157

Adler J, Del Veccio V (1979) Specialized assays for detection of mycoviruses. In: Viruses and plasmids in fungi. Lemke PA (ed). Marcel Dekker Inc

Adler JP, Bozarth RF, MacKenzie DW, Wood HA (1975) In vivo fluorescent antibody screening for fungal viruses and double-stranded RNA. Abstr 75th Am Soc Microbiol

Ball EM (1974) Serological tests for the identification of plant viruses. The Am Phytopathol Soc Plant Virol Comm, St. Paul, Minn, p 31

Day PR, Dodds JA (1979) Viruses of plant pathogenic fungi. In: Viruses and plasmids of fungi. Lemke PA (ed). Marcel Dekker Inc

Day PR, Dodds JA, Elliston JE, Jaynes RA, Anagnostakis SL (1977) Double-stranded RNA in *Endothia parasitica.* Phytopathology 67: 1393–1396

Del Vecchio VG, Dixon C, Lemke PA (1977) Immunoelectrophoretic detection of double-stranded ribonucleic acid from *Agaricus bisporus.* Exp Mycol 1: 102–106

Derrick KS (1978) Double-stranded RNA is present in extracts of tobacco plants infected with tobacco mosaic virus. Science 199: 538–539

Derrick KS, Brlansky RH (1976) Assay for viruses and mycoplasmas using serologically specific electron microscopy. Phytopathology 66: 815–820

Diener TO, Schneider IR (1968) Virus degradation and nucleic acid release in single-phase phenol systems. Arch Biochem Biophys 124: 401–412

Dodds JA (in press) Double-stranded RNA and virus-like particles in *Endothia parasitica.* Proc Am Chestnut Symp, Morgantown, WV, USA

Field AK, Lampson GP, Tytell AA, Hilleman MR (1972) Demonstration of double-stranded ribonucleic acid in concentrates of RNA viruses. Proc Soc Exp Biol Med 141: 440–444

Francki RIB, Jackson AO (1972) Immunochemical detection of double-stranded ribonucleic acid in leaves of sugar cane infected with Fiji disease virus. Virology 48: 275–277

Franklin RM (1966) Purification and properties of the replicative intermediate of the RNA bacterophage R17. Proc Natl Acad Sci USA 55: 1505–1511

Frick LJ (1976) Purification, partial characterization, and comparison of virus-like particles in isolates of *Gaeumannomyces graminis,* the take-all disease fungus. MS Thesis, Purdue University, W Lafayette, Indiana

Frick LJ, Lister RM (1978) Serotype variability in virus-like particles from *Gaeumannomyces graminis.* Virology 85: 504–517

Hollings M (1978) Mycoviruses: viruses that infect fungi. Adv Virus Res 22: 1–53

Hollings M, Stone OM (1971) Viruses that infect fungi. Annu Rev Phytophathol 9: 93–118

Hooper GR, Wood HA, Myers R, Bozarth RF (1972) Virus-like particles in *Penicillium brevi-compactum* and *P. stoloniferum* hyphae and spores. Phytopathology 62: 823–825

Ikegami M, Francki RIB (1973) Presence of antibodies to double-stranded RNA in sera of rabbits immunized with rice dwarf and maize rough dwarf viruses. Virology 56: 404–406

Jackson AO, Mitchell DM, Siegel A (1971) Replication of tobacco mosaic virus. 1. Isolation and characterization of double-stranded forms of ribonucleic acid. Virology 45: 182–191

Koltin Y, Day PR (1976) Suppression of the killer phenotype in *Ustilago maydis*. Genetics 82: 629–637

Lacour F, Michelson AM, Nahon E (1968) Specific antibodies to polynucleotide complexes and their reaction with nucleic acids: importance of the secondary structure of the antigen. In: Nucleic acids in immunology. Plescia OJ, Braun W (eds). Springer, Berlin, Heidelberg, New York pp 32–46

Lesemann DE, Koenig R (1977) Association of club-shaped virus-like particles with a severe disease of *Agaricus bisporus*. Phytopathol Z 89: 161–169

Moffitt EM (1973) Development and application of a serological screening test for the detection of dsRNA mycoviruses. MS Thesis, Purdue University, W. Lafayette, Indiana

Moffitt EM, Lister RM (1973) Detection of mycoviruses using antiserum specific for dsRNA. Virology 52: 301–304

Moffitt EM, Lister RM (1975) Application of a serological screening test for detecting double-stranded RNA mycoviruses. Phytopathology 65: 851–859

Mori K, Kuida K (1977) Virus-like particles in several mushrooms. Abstr 2nd Int Mycol Congr, p 451

Morris TJ (1977) Isolation of double-stranded RNA from virus infected plants and fungi. Proc Am Phytol Soc 4: 160–161

Moyer JW, Smith SH (1976) Partial purification and antiserum production to the 19 x 50 nm mushroom virus particle. Phytopathology 66: 1260–1261

Moyer JW, Smith SH (1977) Purification and serological detection of mushroom virus-like particles. Phytopathology 67: 1207–1210

Ohlson GB, Adler JP, Rose S (1977) Screening for double-stranded RNA by indirect agglutination. Abstr Proc 2nd Int Mycol Congr, p 488

Plescia OJ, Braun W, Imperato S, Cora-Block E, Jaroskova L, Schimbor C (1968) Methylated bovine serum albumin as a carrier for oligo- and poly-nucleotides. In: Nucleic acids in immunology. Plescia OJ, Braun W (eds). Springer, Berlin, Heidelberg, New York, pp 5–17

Plescia OJ, Stramp A, Kwiatowski Z (1969) Hybridization between polyuridylate and oligoadenylate: an immunochemical analysis. Abstr Fed Proc Fed Am Soc Exp Biol 28: 695

Ralph RK, Berquist PL (1967) Separation of viruses into components. In: Methods in virology, vol 2. Maramarosch K, Braun W (eds). Academic Press, New York, pp 463–545

Rose SA, Adler JP (1977) The presence of dsRNA in the dimorphic fungus *Histoplasma capsulatum*. Abstr Proc 2nd Int Mycol Congr, p 575

Sanderlin RS, Ghabrial SA (1978) Physicochemical properties of two distinct types of virus-like particles from *Helminthosporium victoriae*. Virology 87: 142–151

Schur PH, Monroe M (1969) Antibodies to ribonucleic acid in systemic lupus erythematosus. Proc Natl Acad Sci USA 63: 1108–1112

Schwartz EF, Stollar BD (1969) Antibodies to polyadenylate-polyuridilate copolymers as reagents for double-stranded RNA and DNA-RNA hybrid complexes. Biochem Biophys Res Commun 35: 115–120

Voller A, Bidwell D, Huidt G, Engvall EA (1974) A microplate method of enzyme-linked immunosorbent assay and its application to malaria. Bull WHO 51: 209–211

Taxonomy

Taxonomy of Fungal Viruses

M. HOLLINGS

Glasshouse Crops Research Institute, Littlehampton, Sussex BN16 3PU/United Kingdom

1 Virus Classification: a Brief History and the Present Position

Virus taxonomy is a new subject; a satisfactory scheme of virus classification has been developed only in the last 12 years. Many earlier attempts had merely arranged viruses according to host symptoms or type of disease, with some attention to the kind of invertebrate vector, and sometimes to tissue tropisms.

At the International Congress of Microbiology in 1966, the International Committee on Nomenclature of Viruses was formed, with different sub-committees responsible for viruses of vertebrates, invertebrates, bacteria and higher plants (Angiospermae). The objectives were to define groups or genera of viruses, with a type member for each genus, and to test the taxonomic value of the virus cryptogram (Gibbs et al., 1966). This approach differed from all previous attempts in two important respects: firstly, small groups of viruses that shared many properties would be set up; and secondly, the properties used were almost entirely biochemical and biophysical characters, based upon the virus cryptograms. Only viruses that had been adequately characterised were considered, and there was no attempt to force every virus into some group or other. Viruses that did not appear to fit into any of the recognised groups were left for future consideration.

In the following years, many more virus groups were recognised, and the number of "unclassified" viruses correspondingly decreased.

The cryptogram comprised four pairs of symbols, and gave the following information:

1st Pair: Type of nucleic acid/strandedness of nucleic acid. Thus, R/1 = single-stranded RNA.

2nd Pair: M.W. of nucleic acid (in millions)/percentage of nucleic acid in the infective virus.

3rd Pair: Outline of the virus particle (with envelope or occluding protein)/outline of nucleocapsid (nucleic acid + protein most closely associated with it). Thus, S = isometric or spherical; E = elongated, ends not rounded, etc.

4th Pair: Kind(s) of host infected/mode of transmission/kind(s) of vector (if known).

Several important parameters are not given in the cryptogram: the virion size, and its sedimentation coefficient; and the M.W. of the capsid protein(s), although these properties have been essential in defining many of the taxonomic groups.

In the 1976 report of the International Committee on Taxonomy of Viruses (as the name of this body had become) (Fenner, 1976), some 45 families or equivalent major groups of viruses had been recognised and approved by the International Congress for Virology and by the I.C.T.V., and a number of more recently proposed families are currently awaiting approval. In 1975, the Fungus Virus Sub-Committee was formed, and I was asked to be chairman.

2 Classification of Fungal Viruses: the Starting Point

The large majority of reported mycoviruses are, in fact, only viruslike particles (VLP), known from electron micrographs. Some are assumed to contain dsRNA because they induce an interferon response in animals, but adequate data are known for only about a dozen mycoviruses; even for many of these, the full cryptogram cannot be written (Table 1).

Table 1. Cryptograms of isometric dsRNA mycoviruses

Aspergillus foetidus AfV-S:	$R/2: \dfrac{2.24}{*} + \dfrac{2.76}{*}$:S/S:F/C	Ratti and Buck, 1972
Aspergillus foetidus AfV-F:	$R/2: \dfrac{2.31}{*} + \dfrac{1.87}{*} + \dfrac{1.70}{*} + \dfrac{1.44}{*}$:S/S:F/C	Ratti and Buck, 1972
Helminthosporium maydis:	$R/2: \dfrac{6.3}{*}$:S/S:F/*	Bozarth, 1977
Mushroom virus 1:	$R/2: \dfrac{2.6}{*} + \dfrac{1.6}{*} + \dfrac{0.26}{*} + \dfrac{0.23}{*}$:S/S:F/C	Atkey et al., 1975
Mushroom virus 4:	$R/2: \dfrac{1.5}{*} + \dfrac{1.5}{*}$:S/S:F/C	Atkey et al., 1975
Penicillium chrysogenum:	$R/2: \dfrac{2.18}{11} + \dfrac{1.99}{11} + \dfrac{1.89}{11}$:S/S:F/C	Wood and Bozarth, 1972 Border et al., 1972
P. stoloniferum PsV-S:	$R/2: \dfrac{1.01}{24} + \dfrac{0.94}{24}$:S/S:F/C	Bozarth et al., 1971
P. stoloniferum PsV-F:	$R/2: \dfrac{0.99}{*} + \dfrac{0.89}{*} + \dfrac{0.23}{*}$:S/S:F/C	Buck and Kempson-Jones, 1973
Periconia circinata:	$R/2: \dfrac{1.75}{*} + \dfrac{1.40}{*} + \dfrac{1.25}{*} + \dfrac{1.10}{*} + \dfrac{0.48}{*} + \dfrac{0.42}{*}$:S/S:F/*	Dunkle, 1974
Colletotrichum lindemuthianum:	$R/2: \dfrac{2.5}{*} + \dfrac{1.1}{*} + \dfrac{1.0}{*}$:S/S:F/*	Rawlinson et al., 1975

In this rather discouraging situation, the best approach seemed firstly, to see what we can learn from the taxonomic methodology of our colleagues dealing with viruses in other host taxa; secondly, to evaluate what seem to be the "minimum essential properties" that need to be known about a virus before it can be considered for taxonomic grouping; and thirdly, to ask our colleagues working with mycoviruses to cooperate by determining these properties for their own viruses.

3 Comparative Survey of Viruses in Other Host Taxa

If the recognised virus groups are arranged according to their genome composition and host taxon, it can be seen that four genome types (ssDNA, dsDNA etc.) occur in all host taxa, except the fungi, for which information is incomplete (Table 2). However, if the numbers of individual viruses are tabulated in a similar manner (Table 3), some

Table 2. Families or groups of viruses, according to genome composition

Host taxa	ssDNA	dsDNA	ssRNA	dsRNA
Vertebrates	1	5	9	1
Invertebrates	1	2	2	1
Plants	1	1	21	1
Bacteria	2	4	1	1
Fungi	?	?1	1	>1
Total	5	13	34	5

Table 3. Numbers of viruses, according to genome composition

Host taxa	ssDNA	dsDNA	ssRNA	dsRNA
Vertebrates	19	122	300	16
Invertebrates	2	49	64	1
Plants	10	6	214	7
Bacteria	34	82	18	1
Fungi	?	?1	1	>20
Total	65	260	597	45

interesting features are seen. Most of the known dsRNA viruses occur in fungi, and the figure "more than 20" is likely to be greatly exceeded when more VLP have been characterised.

By contrast, ssRNA viruses occur predominantly in the vertebrates and especially in higher plants, as reflected both in the number of taxonomic groups present in these hosts, and in the number of different viruses involved (Tables 2, 3). If we regard the host taxon as the habitat or ecological niche, it appears that ssRNA viruses are well suited to vertebrate and higher plant cells, and these viruses show a wide diversity, as reflected in the many different groups of families. Likewise, present evidence suggests that, for some reason, isometric dsRNA viruses are particularly well adapted to flourish in fungus cells. From the taxonomic aspect, there is no a priori reason to assume that all dsRNA mycoviruses belong to only one group, merely because they have a seg-mented dsRNA genome.

Table 4. Criteria used in grouping ssRNA multicomponent plant viruses

Virus group	Virion		RNA genome			Capsid protein		Vectors
	Diam. (nm)	$s_{20,w}$	No. of pieces	M.W. ($\times 10^6$)	% in virion	M.W. ($\times 10^3$)	Serological relationships among members	
Bromovirus	25	85 S	4	1.1 1.0 0.8 0.3	22	20	Some	(Beetle)
Comovirus	30	115 S 95 S 55 S	2	2.3 1.4 —	34 23 0	22 and 42	Yes	Beetle
Cucumovirus	30	98 S	4	1.1 1.0 0.7 0.3	18	24	Yes	Aphid (stylet)
Ilarvirus	26–35	80–110 S	4	1.3 1.1 0.8 0.4	14	27	Yes (2 sub-groups)	(Pollen)
Machlovirus	30	154 S	1	3.2	36	?	?	Leafhopper
Nepovirus	30	125–128 S 95–119 S	3–4	1.2 2.3 —	43 27 0	55	No	Nematode
Penamovirus	28	115 S 90 S	2	1.7 1.4	29	22	(Monotypic)	Aphid (circulative)

Table 5. Criteria used in grouping isometric ssRNA single-component plant viruses

Virus group	Virion		RNA genome			Capsid protein		Vectors
	Diam. (nm)	$s_{20,w}$	No. of pieces	M.W. ($\times 10^6$)	% in virion	M.W. ($\times 10^3$)	Serological relationships among members	
Sobemovirus	28	115 S	1	1.4	21	30	No	(Beetle)
Tobanecvirus	28	118 S	1	1.5	19	23	Yes	Chytrid fungus
Tombusvirus	30	140 S	1	1.5	17	41	Mostly yes	Soil
Tymovirus	30	110 S 50 S	1	2.0 –	36 0	20	Yes	Beetle
(Carnation mottle virus group) [a]	28	122 S	1	1.4	20	38	No	Contact
(Phleum mottle virus group) [a]	28	106 S	1	1.5	?	25	Yes	Beetle

[a] Groups proposed unofficially, and not recognised by I.C.T.V. (Hull, 1977)

4 Taxonomic Grouping of the Isometric ssRNA Plant Viruses

Let us examine the grouping of isometric ssRNA plant viruses, especially those with segmented genomes, and consider the importance and usefulness of the different in vitro properties used to classify them (Table 4). Virion size is of little help in separation, for all these viruses (except the Ilarvirus group) have particles between 25 and 30 nm diameter. Virion sedimentation coefficients are much more useful, and are characteristic for each virus group. The number of RNA genome segments and their molecular weights are interesting: firstly, because these values separate the Comovirus, Nepovirus and Penamovirus groups from the others; and secondly because they suggest possible affinities between the Bromovirus, Cucumovirus, and Ilarvirus groups. However, the wide differences in the percent nucleic acid in the virion emphasises the differences between these groups, and this is supported by the differences in the M.W. of the capsid protein. Finally, there are the serological relationships among member-viruses within each of the groups, except the Nepoviruses.

If we now consider the isometric ssRNA plant viruses with a single genome segment (Table 5), again we find that virus particle diameter is of little use in separating virus groups. Sedimentation coefficients are of some value, especially at the higher and lower ends of the range of values. The M.W. of the RNA genome is of limited use, with figures around $1.4-1.5 \times 10^6$ daltons for most of these viruses. The percentage RNA in the virion is of more value, but the most useful parameter seems to be the M.W. of the capsid protein, and the different virus groups have characteristic values for this. Biological properties, such as vector type and relationships, confirm the separations made on in vitro properties, and there are serological relationships among the members in some of the groups. Whatever theoretical criticisms can be made of serological relationships as taxonomic criteria, in practice serology has proved sound and reliable.

It is perhaps worth mentioning some of the virus properties that, for various reasons, have *not* been of direct use in separating taxonomic groups: buoyant density in CsCl or Cs_2SO_4; sedimentation coefficient(s) of the RNA genome segments; nucleotide base ratios (except in the Tymovirus group); and nucleic acid hybridisation tests. Some of these properties, however, have provided additional confirmation to support the validity of the proposed virus groups.

5 Relative Importance of Different Properties for Grouping Mycoviruses

Any assessment of the importance of a virus property is largely a personal and subjective one; an Adansonian approach, giving all properties equal importance, is also open to criticism, for some properties are inevitably correlated with others. For example, sedimentation coefficients and buoyant density values will depend largely on the particle size and the amount and molecular weight of nucleic acid. The views of the I.C.T.V. Fungus Virus Sub-Committee on the relative values of different properties are given in Table 6. This is a formidable list, and it is very unlikely that all these properties will ever be determined for all the mycoviruses, simply because time and research programs do not permit this. For taxonomic purposes, however, some properties are much more *useful* than others, because there are much greater differences

Table 6. Relative taxonomic importance of various properties of mycoviruses: consensus opinion of I.C.T.V. Fungus Virus Sub-Committee

Total points gained (out of 40)	Criteria
40	Nucleic acid – DNA/RNA Nucleic acid – ss/ds
39	Virion shape and dimensions in electron micrographs Serological relationships (when *positive*)
38	Nucleic acid – number of pieces Nucleic acid – molecular weights Nucleic acid – circular, linear or looped
36	Nucleic acid – mode of encapsidation (into several particles, or 1) Coat protein – no. of species
35	Virion sedimentation coefficient(s)
34	Coat protein – molecular weight(s)
33	Percent nucleic acid in virion Percent protein in virion Symmetry of virion
31	Nucleic acid hybridization (where *positive*) Serological reactions (where negative)
29	Virion molecular weight Presence and percent of lipid in virion
28	Presence of internal protein(s) (including enzymes?)
24	Buoyant density in CsCl and Cs_2SO_4
23	Triangulation number Nucleic acid hybridization (where negative)
22	Nucleic acid sedimentation coefficient(s)
18	Virion sensitivity to EDTA
17	Host range – genera or groups of fungi infected
16	Virion sensitivity to ether
14	Virion sensitivity to chloroform Virion sensitivity to pH Virion sensitivity to SDS
13	Virion sensitivity to proteases Formation of intracellular crystals or inclusions
12	Sensitivity of virion to heat Sensitivity of virion to urea Electrophoretic migration of virion
11	Sensitivity of virion to citrate
10	Latent period in host
7	Sensitivity of virion to formaldehyde Sensitivity of virion to freezing/thawing Sensitivity of virion to osmotic shock Sensitivity of virion to sonication Sensitivity of virion to U.V. irradiation
6	Cytopathic effects

between virus groups than within groups. Moreover, some properties are valuable for differentiating certain virus groups, but are of little use with other kinds of viruses. A list of "minimum essential information" can be selected from among these properties, and this presents a much less formidable barrier (Table 7).

Table 7. "Minimum essential information" for mycovirus taxonomy

I. Virion:	Shape and dimension in electron microscope: stain and calibration standard to be specified
	Symmetry: cubic, helical, etc.
	Sedimentation coefficient(s): correction factors to be specified
	Serological reactions: tests for relatedness to other mycoviruses
II. Nucleic acid:	RNA or DNA
	ss or ds
	Number of pieces
	Molecular weights of pieces
	Percent NA in virion
	Mode of encapsidation: NA segments in 1 or several particle types
III. Protein:	Number of coat proteins
	Molecular weights of coat proteins
	Percent in virion
IV. Lipid:	Presence in virion
	Percent in virion

This is where we need the help and co-operation of colleagues working with mycoviruses, to determine these properties for the viruses that they are studying. Even for this minimum list of properties, the data are so far known for only one or two mycoviruses, and partially complete data for another ten. For several viruses, there are surprising discrepancies between the properties determined in different laboratories; for example, the reported particle diameters in electron microscopy differ by as much as 20% for several mycoviruses. Standardisation of procedures is therefore essential.

6 Taxonomic Grouping of dsRNA Isometric Mycoviruses

Even from the limited available information, there is clearly a wide range of properties among the dsRNA mycoviruses: virions range from 25–50 nm in diameter, and from $6-13 \times 10^6$ in M.W. The single-layered capsids have structural proteins of M.W. from $42-130 \times 10^3$, and the 2–5 pieces of dsRNA range in M.W. from $0.24-2.76 \times 10^6$, with totals of $2.04-8.56 \times 10^6$ in different viruses. This diversity suggests the existence of a number of separate taxonomic groups among these viruses.

As a starting point, we have provisionally suggested the creation of two groups of isometric dsRNA mycoviruses: the *Penicillium chrysogenum* virus group, and the *P. stoloniferum* PsV-S virus group. The details are defined in Table 8. Within each group, the member viruses are serologically related.

Table 8. Proposed first taxonomic groups of dsRNA mycoviruses

---

1. *Penicillium chrysogenum* virus group

Type member:	*P. chrysogenum* virus
Cryptogram:	$R/2 : \dfrac{2.18}{11} + \dfrac{1.99}{11} + \dfrac{1.89}{11} : S/S : F/C$
Virion:	Icosahedral, 35–40 nm diameter, $S_{20,w} = 150$ S, wt $= 13.0 \times 10^6$, buoyant density 1.354 g/ml in CsCl Serologically related to closely similar viruses of *P. brevicompactum* and *P. cyaneo-fulvum*
Nucleic acid:	3 pieces of dsRNA (4 in *P. cyaneo-fulvum*), linear, separately encapsidated, and comprising ca. 11% of particle wt. G + C = 56%
Protein:	Single shell, 1 capsid protein, ca. 89% of particle wt. RNA polymerase present
Lipid:	None
Carbohydrate:	None
Occurrence:	Cytoplasm
Pathogenicity:	None reported
Transmission:	Congenital through conidia; by hyphal anastomosis
Other members:	Viruses of *P. brevicompactum* and *P. cyaneo-fulvum*

2. *Penicillium stoloniferum* PsV-S group

Type member:	PsV-S
Cryptogram:	$R/2 : \dfrac{1.01}{24} + \dfrac{0.94}{24} : S/S : F/C$
Virion:	Icosahedral, ca. 34 nm diameter, $S_{20,w} = 113, 101, 87$ and 66 S.Wt. $= 5.04 – 6.67 \times 10^6$, buoyant density 1.3–1.38 g/ml in CsCl Serological relationships: related to virus from *Diplocarpon rosae*
Nucleic acid:	2 pieces, linear, dsRNA, separately encapsidated comprising ca. 24% of particle weight. ssRNAs of half the ds molecular weights sometimes also found
Protein:	Single shell, 2 capsid structural proteins, 42.5 and 55.5, comprising ca. 76% of particle wt. RNA polymerase present
Lipid:	None
Carbohydrate:	None
Occurrence:	Cytoplasm
Pathogenicity:	None reported
Transmission:	Congenital through conidia; by hyphal anastomosis
Other members:	Virus of *Diplocarpon rosae*

---

7 Other Mycoviruses

So few of the properties in vitro are known for the rod-shaped mycoviruses (or VLP), that no meaningful attempts can yet be made to classify them. From particle morphology, however, it is possible that at least some of these may be candidates for membership of the Tobamovirus group.

The bacilliform particles of mushroom virus 3 (MV-3, 19 x 50 nm) show obvious morphological similarities to those of alfalfa mosaic virus (AMV), but the two viruses are serologically unrelated. Like AMV, MV-3 can re-aggregate to make long tubular "nonsense assemblies" from partly disrupted virions. Both viruses contain ssRNA, and MV-3 has one capsid protein of M.W. ca. 16×10^3. The optical diffraction patterns from micrographs of MV-3 particles are very like those of AMV, with lattice spacings suggesting a sub-unit ca. 10 nm diameter in MV-3 and 9.6 nm in AMV.

The large, DNA-containing virus from *Thraustochytrium* sp. is very similar to typical Herpesviruses in particle morphology and in what is known of its replicative cycle. More hard data are required before it could be definitely assigned to the Herpetoviridae.

A number of other large VLP have been reported, of various shapes and sizes, but insufficient data are known about their properties in vitro for any taxonomic considerations to be possible. These have been further discussed elsewhere (Hollings, 1978).

Summary

In the past 12 years, an acceptable taxonomic system has been produced for viruses of vertebrates, angiosperms and bacteria. Fewer than 10% of the reported viruses and VLP of fungi, however, have been adequately characterised for taxonomic purposes.

In comparing isometric mycoviruses with those in other host taxa, the dsRNA genome apparently predominates in fungal viruses, whereas ssRNA viruses are the most diverse and successful groups in vertebrates and plants. Available evidence suggests that dsRNA mycoviruses may form a number of separate taxonomic groups; two such groups are provisionally proposed: (1) *Penicillium chrysogenum* virus group; and (2) the *P. stoloniferum* PsV-S virus group. Within each group, the member viruses are serologically related.

A list of "minimum essential information" required for each mycovirus has been suggested, and all mycovirus workers are asked to help to determine these properties for the viruses they study.

References

Atkey PT, Barton RJ, Hollings M, Stone OM (1975) Mushroom virus purification and characterization. Rep Glasshouse Crops Res Inst 1974, p 121

Border DJ, Buck KW, Chain EB, Kempson-Jones GF, Lhoas P, Ratti G (1972) Viruses of *Penicillium* and *Aspergillus* species. Biochem J 127:4–6

Bozarth RF (1977) Biophysical and biochemical characterization of virus-like particles containing a high molecular weight ds-RNA from *Helminthosporium maydis*. Virology 80: 149–157

Bozarth RF, Wood HA, Mandelbrot A (1971) The *Penicillium stoloniferum* virus complex: Two similar double-stranded RNA virus-like particles in a single cell. Virology 45: 516–523

Buck KW, Kempson-Jones GF (1973) Biophysical properties of *Penicillium stoloniferum* virus S. J Gen Virol 18: 223–235

Dunkle LD (1974) Double-stranded RNA mycovirus in *Periconia circinata*. Physiol Plant Pathol 4: 107–116

Fenner F (1976) Classification and nomenclature of viruses. Intervirology 7: 1–115

Gibbs AJ, Harrison BD, Watson DH, Wildy P (1966) What's in a virus name? Nature (London) 209: 450–454

Hollings M (1978) Mycoviruses – viruses that infect fungi. Adv Virus Res 22: 1–53

Hull R (1977) The grouping of small spherical plant viruses with single RNA components. J Gen Virol 36: 289–295

Ratti G, Buck KW (1972) Virus particles in *Aspergillus foetidus:* a multicomponent system. J Gen
   Virol 14: 165–175
Rawlinson CJ, Carpenter JM, Muthyalu G (1975) Double-stranded RNA virus in *Colletotrichum
   lindemuthianum.* Trans Br Mycol Soc 65: 305–308
Wood HA, Bozarth RF (1972) Properties of viruslike particles of *Penicillium chrysogenum:* one
   double-stranded RNA molecule per particle. Virology 47: 604–609

Part B
Symposium on Extrachromosomal Vectors in Fungi – Abstracts

Chairman: E. A. BEVAN
Convenor: K. ESSER

Introductory Remarks of the Chairman

E.A. BEVAN

Dept. of Plant Biology and Microbiology, Queen Mary College, University of London/
United Kingdom

The title of this symposium has been purposely chosen so that we may gain some insight into the progress which has been made (a) in establishing the in vivo role of cytoplasmically inherited nucleic acid molecules in fungi, and (b) evolving techniques which permit their isolation, manipulation in vitro, and reinfection into host cells.

The knowledge already gained concerning the molecular structure and behaviour of some of the extrachromosomal elements of fungi gives some confidence that further investigations may lead to their uses as vectors of "foreign" DNA, in the same way as has been accomplished using plasmids and bacteriophages of bacteria. It is now becoming increasingly apparent that eukaryotic genes transferred to bacteria, with very few exceptions (e.g., yeast genes), produce a different gene product to that produced in their "home" environment. This is not surprising in view of our relative ignorance of gene regulatory mechanisms in eukaryotes, and the known differences in protein synthesis between prokaryotes and eukaryotes. Clearly, the aim of using eukaryotic fungi instead of prokaryotic bacteria for gene cloning purposes is one worth pursuing, since there is every reason to believe that their use may ultimately lead to the realisation of both the fundamental scientific and economic benefits originally predicted as being possible using the plasmid/phage bacterial system. Two recent discoveries worthy of note have been made since this symposium was originally conceived, and it is hoped that the implications of both will be discussed by participants during the symposium. The first of these is the success of Hinner, Hicks, and Fink in demonstrating that auxotrophic strains of yeast *(Saccharomyces cerevisiae)* can be transformed to prototrophy by introduction of their protoplasts of ColE1 bacterial plasmids carrying the leu_{10} gene of yeast. DNA hybridization and genetic analyses have confirmed that in the majority of prototrophic cells produced in this way both the leu_{10} yeast gene and ColE1 plasmid DNA becomes incorporated into the chromosomes.

The second discovery worthy of discussion is that made by Stahl, Lemke, Tudzynski and Esser. These workers have recently demonstrated the existence of a plasmid-like DNA element in the filamentous ascomycete fungus *Podospora anserina*. The presence of such a plasmid-like DNA in a filamentous fungus offers a new "tool" for both basic and applied research and provides a further eukaryotic system which may ultimately prove useful in gene cloning programmes.

Recent Developments on the Genetics of the URE 3 Cytoplasmic Factor

M. AIGLE and F. LACROUTE

I.B.M.C du C.N.R.S., 15 rue R. Descartes, 67084 Strasbourg Cedex/France

It has been previously established that an extrachromosomal factor named URE 3 which is characterized by its non-Mendelian inheritance was not linked to the mitochondrial genome and was incompatible with a chromosomal mutation URE 2 displaying the same phenotype. We have studied more recently the possible relationship of the URE 3 factor with other cytoplasmic elements described in yeast. There is no genetic interaction with the ψ factor described by Cox. There is no correlation between the URE 3 phenotype and the killer double-stranded RNA; moreover the URE 3 phenotype is not cured by ethidium bromide, cycloheximide or heat. The restriction profile of $2\,\mu$ DNA in URE 3 and wild-type strains is the same (which does not discard the possibility of $2\,\mu$ being the vector). There is a possible relationship which is presently being studied with the cytoplasmic heredity of a 20S RNA found by Haber. The recent improvements in yeast transformation are used to try to identify the URE 3 vector.

The Organization of the Mitochondrial Genome of Saccharomyces cerevisiae and Its Implications on the Phenomenon of Suppressiveness

H. BLANC

Centre de Genetique Moleculaire, C.N.R.S., Gif sur Yvette/France

The mitochondrial genome of respiratory competent strains, ρ^+, of *S. cerevisiae* consists of circular DNA molecules of around 75,000 bp. This molecule specifies the two rRNA and the tRNA components of the mitochondrial protein-synthesizing system, and genes for a number of inner membrane proteins. A precise physical map of the mtDNA is now available, on which the different genetic loci have been located, using mainly the petite mutation (ρ^-). This mutation consists of large deletions of the ρ^+ DNA molecule, compensated by the reiteration of the nondeleted fragment in such a way that the maintained segment is highly amplified. The physical maps of different ρ^+ strains, all wild types but of different origins, show similar general organization but differ by the presence or absence of several DNA segments inserted at specific sites along the molecule. In this presentation two of the problems currently investigated in our laboratory will be discussed. They concern split genes and suppressiveness.

A specific insert (1000 bp) was located within the 23S rRNA gene, splicing it. The presence of this insert is correlated with the ω^+ allele, which is responsible for polarity of recombination. The gene coding for the cytochrome b polypeptide was shown to be split into several exons separated by at least two introns. When a population of ρ^- cells is crossed to a population of ρ^+ cells, a certain proportion of diploids is composed of ρ^- cells. The degree of suppressiveness, that is to say the percentage of the progeny ρ^-, is a hereditary property of each ρ^- clone. Some ρ^- clones can be more than 99% suppressive. Several specific questions about the molecular nature of highly suppressive ρ^- and the mechanism of suppressiveness will be discussed.

The precise mapping of the mtDNA, the capacity of amplification of genes in ρ^-, the high efficiency of recombination, and finally the possibility of transformation of yeast, makes it possible to consider the use of the mtDNA as a potential vector for cloning foreign genes.

Sequence Organization and Expression of a Plasmid DNA Molecule Found in Yeast

J.E. DONELSON, J.L. HARTLEY, E.J. GUBBINS, and D.L. MILLER

Department of Biochemistry, University of Iowa, Iowa City, IA 52242/USA

Most strains of baker's yeast, *Saccharomyces cerevisiae,* contain closed-circular duplex DNA plasmid molecules approximately 2 μm (6,200 base pairs) in contour length. There are about 100 plasmid molecules per cell, the sum of which comprises about 2.5% of the haploid genome. The plasmid possesses a nontandem inverted repeat sequence of about 650 base pairs about which intramolecular reciprocal recombination occurs to generate two forms of sequence organization. We have identified a 750 base pair restriction enzyme fragment that contains the inverted repeat sequence and have determined its complete nucleotide sequence by Maxam/Gilbert analysis. We have also screened a bank of approximately 3,000 random yeast DNA fragments cloned in *E. coli* via the poly dA:T method for the presence of yeast plasmid sequences. The hope was that we would identify a cloned yeast DNA fragment which contained a boundary between the yeast plasmid sequence and a yeast chromosomal sequence. This would prove that the yeast plasmid was capable of integrating into chromosomal DNA. Twelve of the 3,000 clones were found to possess yeast plasmid sequences, but none possessed the hoped-for boundary between the yeast plasmid and chromosomal DNA. Nevertheless, some of the cloned yeast plasmid sequences were dimeric forms which allowed us to conduct experiments which indicated that (1) at least one molecular mechanism of recombination in yeast is different from that in *E. coli* and (2) the intact yeast plasmid monomer is unable to replicate by itself in *E. coli.* Some yeast strains do not possess the plasmid DNA sequences [Livingston, D.M.: Genetics *86*: 73–84 (1977)]. The total protein fingerprint profile of these strains on O'Farrell 2-dimensional gels does not appear different from that of yeast strains which have about 2.5% of their total cellular DNA in the plasmid sequence.

Structure and Function of Saccharomyces cerevisiae 2-μm DNA

C.P. HOLLENBERG and H.D. ROYER

Max-Planck-Institut für Biologie, Abt. Beermann, Tübingen/FRG

Yeast 2-μm DNA is a closed-circular extra-chromosomal DNA element of which 50—100 copies are normally found in several strains of *Saccharomyces cerevisiae*. No gene products of this DNA element are known and a direct function has not yet been established. To obtain information on possible gene products of this molecule, we engaged in a study of the expression of cloned yeast 2-μm DNA in *Escherichia coli*. Integrated at the EcoRI site of pCR1 and pBR313 or at the PstI site of pBR322, the 2-μm DNA promoted the synthesis of polypeptides of 48,000, 37,000, 35,000, and 19,000 molecular weight. The DNA regions coding for these polypeptides were mapped on the 2-μm DNA molecule by insertion of single EcoRI or HindIII restriction fragments. The cloned inverted repeat sequence of yeast 2-μm DNA did not induce any detectable insert-specific polypeptide synthesis.

In connection with the expression of 2-μm DNA in *E. coli,* we have mapped the *E. coli* RNA polymerase binding sites. On native 2-μm DNA from *S. cerevisiae* strain HQ/5C five RNA polymerase binding sites were detected. One further site was mapped on cloned 2-μm DNA type 23 from *S. cerevisiae* strain H1. This additional site is located at a distance of 2.15 kb from EcoRI site B inside one of the inverted duplication (id) sequences. The other id sequence does not contain a binding site. Thus it is clear that the two id sequences of strain H1 are not identical but differ in at least one short region (PB site). The location of the id sequence containing the PB site was analyzed in native 2-μm DNA from strain H1. The ability to identify each id sequence allowed us to gain further insight into the structure of the two types of 2-μm DNA which occur in the yeast cell and the mechanism by which these might arise. The location of the id sequences within 2-μm DNA suggests that only one of the unique segments of the molecule (L-loop region) is invertible between the id sequences. The other unique segment remains linked to at least that part of the id sequence including the PB site. The recombinational event assumed to generate the two types of molecules thus cannot take place over the entire id sequence, but has to be confined to its right arm, probably at the border between id sequences and unique segment L. The mechanism of recombination, therefore, seems to resemble a site-specific event such as found associated with bacterial IS sequences more than a sequence-specific recombination process.

The location of only one *E. coli* polymerase binding site was such that it could serve as a promoter for one of the 2-μm DNA products.

Elements of Infectious Heredity in Fungi

P.A. LEMKE

Mellon Institute, Carnegie-Mellon University, Pittsburgh, PA 15213/USA

In its broadest and perhaps original sense the term "plasmid" encompasses a cytoplasmically inherited genetic supplement to a cell's normal genetic complement. In the fungal cell, where eukaryotic cellular organization is manifested, the normal genetic complement includes nuclear and mitochondrial DNA. All other informational molecules of nucleic acid resident in fungal cytoplasm, if they can be inherited, should qualify as plasmids and, in the context of this symposium, as potential extrachromosomal vectors.

What are these molecules and how are they inherited? The evidence extant indicates that fungal plasmids include double-stranded RNA (dsRNA) and covalently-closed-circular DNA (cccDNA), and both types of molecules are infectious through heredity or cytoplasmic exchange. This exchange and, to some extent, the replication and expression of fungal plasmids, is influenced by nuclear genes. Among such nuclear genes are loci for either vegetative or sexual incompatibility, insofar as these genes influence the extent of cytoplasmic exchange among fungal populations.

The dsRNA plasmids of fungi are associated principally with isometric viruslike particles. Populations of virus particles have been shown to be multicomponent for genome segments of dsRNA. Particles per se have not been shown to be infectious, and specific functions for only a few dsRNA segments have been determined in just two fungal systems.

With regard to cccDNA, only the *Saccharomyces cerevisiae* plasmid has been physicochemically well characterized. Physical evidence for another DNA plasmid has been obtained in *Podospora anserina*. This plasmid can now be investigated for function among certain cytoplasmically inherited phenotypes of this filamentous fungus.

Part C
Posters on Fungal Viruses – Abstracts

Mycoviruses in Helminthosporium

A. MISRA[1], T.K.S. SINGH[2], Y.P. YADAV[3], B. MISHRA[3], and D.P. CHOUDHARY[4]

[1] Botany Department, Mithila University, Darbhanga 846 004/India. [2] Botany Department, Bihar University, Muzaffarpur/India. [3] Department of Plant Pathology, Rajendra Agriculture University, Dholi, Bihar/India. [4] Botany Department, C.M. Sc. College, Darbhanga/India

Fungal viruses are well known by now, and have been reported in *Helminthosporium* species from several laboratories. However, their occurrence in our country is less known. We hereby report the presence of mycoviruses in *Helminthosporium*. Cultures of the fungus showed watery, hyaline spots which were able to be transmitted by physical transfer to further cultures. This abnormality also seemed to be inherently present in some of the forms, while totally absent from some others. Gramineous *Helmintho-sporia* are more prone to mycoviruses than non-graminicolous forms. Morphology, transmission, and ultrastructure of the viruses will be presented, and taxonomy of mycoviruses in general will be discussed.

Viruses of Aspergillus awamori and the Production of Glucoamylase

L.J. VITALE, R. VALINGER, D. VLAŠIĆ, and J. BELJAK

Institute "Rudjer Bošković", Zagreb, Yugoslavia, "Pliva" Pharmaceutical and Chemical Works, Zagreb/Yugoslavia

Many fungi, industrial producers of enzymes and antibiotics, are known to carry viruses, but little is known about effects of viruses on product biosynthesis. We have examined *Aspergillus awamori,* a glucoamylase-producing strain, for the presence of viruses and the correlation between the virus content and enzyme production.

The mould extracts were concentrated by PEG precipitation and precipitates fractionated by differential centrifugation. The sediment obtained at high speed contained viruses as revealed by electron microscopy. The existence of two virus types, also known in other *Aspergillus* strains, was shown by electrophoretic and immunoelectrophoretic analysis. Further purification and separation of two virus types was achieved by gel filtration on Sepharose 4B and by DEAE cellulose chromatography. For quantitative determination of viruses, specific immune sera were prepared and used in an immunoelectrophoretic assay which was adapted for virus analysis. Virus concentration was determined in mycelia of *A. awamori* parent strain and strains obtained after temperature treatment, and was correlated with strain ability to produce glucoamylase. The results indicate that there is a certain correlation between virus titer and enzyme production.

Sporulation and dsRNA Viral Molecules in Penicillium citrinum

L. VOLTERRA, R. BENIGNI, E. DOGLIOTTI, and G. IGNAZZITTO

Instituto Superiore die Sanità, Viale Regina Elena, 299 00161 Rome/Italy

Viral distribution and sporulation seemed to be characters related to each other in a strain of *Penicillium citrinum*. While the sporulating isolate exhibited 10 dsRNA molecules whose molecular weights were respectively: 3.98, 2.66, 2.48, 2.10, 1.97, 1.84, 1.78, 1.40, 1.26, and 1.10×10^6 daltons, the nonsporulating one contained only 3 dsRNA molecule species with the following molecular weights: 3.98, 1.26 and 1.10×10^6 daltons. We supposed that only these two last dsRNA molecules were responsible for the nonsporulation of the isolate. In order to show better the correlation between sporulation and the presence of particular viral molecules, we treated the nonsporulating isolate with several salts at various concentrations. Initially we used only $MnCl_2$ to induce sporulation. It was known to induce morphological variations in strains of *Aspergillus nidulans*, that appeared very close to those described in *Aspergillus glaucus* and could be explained as probably due to a mycoviral infection. After the first encouraging results we extended our experiments to other salts. We found that NiCl, $FeCl_3$, $MnCl_2$, RbCl, and $BaCl_2$ were good inducers of sporulation. The experiments of induction were made on Czapek Dox medium plus the substances under test. Fragments of nonsporulating mycelium were placed in the middle of petri dishes. After 15—20 days of incubation at 24°C, we selected the green sporulating sectors. These were used to inoculate flasks containing potato dextrose broth, pH = 7.0. After 5 days of fermentation at 24°C, the mycelia were harvested and virus extracted. Viral nucleic acid extracts were obtained directly from the virus preparations and compared by 2.6% polyacrylamide gel electrophoresis. We found that all strains in which sporulation was induced were characterized by the constant presence of 2.66, 2.10, and 1.97 $\times 10^6$ daltons dsRNA molecular species; therefore we suppose that among the 10 dsRNA molecules just these three were particularly connected with the appearance of the sporulation.

Subject Index

Page numbers in *italics* refer to illustrations

R. A. Robinson

Plant Pathosystems

1976. 15 figures, 2 tables. X, 184 pages
(Advanced Series in Agricultural Sciences,
Volume 3)
ISBN 3-540-07712-X
Distribution rights for India: Allied Publishers
Private Ltd., New Delhi

Contents:
Systems. – Plant Pathosystems. – Vertical
Pathosystem Analysis. – Vertical Pathosystem
Management. – Horizontal Pathosystem Ana-
lysis. – Horizontal Pathosystem Management. –
Polyphyletic Pathosystems. – Crop Vulnerabi-
lity. – Conclusions. – Terminology.

"… because *Plant Pathosystems* is already a very
good book and unique in many ways. The text
ist based on long and varied experience of field
plant pathology in Africa and essentially is a
series of perceptive analyses of interactions between populations of host plants and populations of pathogens, … The book is well and
clearly written but must be taken slowly because
each paragraph on each page contains one or
more substantial points which must be understood if the main themes which are propounded
are to have their full impact. But the effort called
for is very well rewarded because few books published on plant pathology during the past three
decades are so stimulating, challenging, and
have been written with such worthy objectives.
Plant Pathosystems will undoubtedly cause
quite a stir among plant pathologists and plant
breeders, for most of whom it must be regarded
as compulsory reading."

Nature

Springer-Verlag
Berlin
Heidelberg
New York

H. C. Coppel, J. W. Mertins

Biological Insect Pest Suppression

1977. 46 figures, 1 table. XIII, 314 pages
(Advanced Series in Agricultural Sciences,
Volume 4)
ISBN 3-540-07931-9
Distribution rights for India: Allied Publishers
Private Ltd., New Delhi

Contents:
Glossary. – Historical, Theoretical, and Philo-
sophical Bases of Biological Insect Pest Suppression. – Organisms Used in Classical Biological
Insect Pest Suppression. – Manipulation of the
Biological Environment for Insect Pest Suppression. – A Fusion of Ideas. – References. – Index.

This cohesive review of many areas of entomo-
logy describes the potential and practical aspects
of suppressing insect pest populations by a
variety of biological methods. These include the
introduction and encouragement of natural ene-
mies, host resistance, hormones, pheromones,
antifeedants, genetics, and integrated systems.

Microbial Ecology

Editors: M. W. Loutit, J. A. R. Miles
1978. 173 figures, 123 tables. XXII, 452 pages
(Proceedings in Liefe Sciences)
ISBN 3-540-08974-8
Distribution rights for India: Narosa Publishing
House, New Delhi

Contents:
Microbial Ecology – General. – Marine Micro-
bial Ecology. – Freshwater Microbial Ecology. –
Biodegradation in Aquatic Environments. –
Microbial Ecology of Soils. – Microbial Ecology
of Animals: Microorganisms and the Gastro
Intestinal Tract. Microbial Insecticides. –
Microbial Ecolgoy of Plants: The Rhizosphere.
Plant Diseases. Rhizobium – Legume Symbi-
osis. – Environmental Problems Management
and Control. – Microbiology of Food.

This book contains selected original papers presented at the First International Symposium on
Microbial Ecology held at the University of
Otago, Dunedin, NZ, August 22–26, 1977. The
papers, presented by scientists from all over the
world, are grouped into sections on the microbial ecology of animals, plants, soil, water, and
food. There are also sections on environmental
problems, their management and control.
Since this volume publishes results of work in
progress it will be of considerable interest to
those actively engaged in microbial ecology.
Papers on all aspects of this field give the book a
wider scope than the majoritiy of available textbooks.

Springer Series in Microbiology

Editor: Mortimer P. Starr

T. D. Brock
Thermophilic Microorganisms and Life at High Temperatures

1978. 195 figures, 69 tables. XI, 465 pages
ISBN 3-540-90309-7

Contents:
The Habitats. – The Organisms: General Overview. – The Genus Thermus. – The Genus Thermoplasma. – The Genus Sulfolobus. – The Genus Chloroflexus. – The Thermophilic Blue-Green Algae. – The Genus Cyanidium. – Life in Boiling Water. – Stromatolites: Yellowstone Analogues. – A Sour World: Life and Death at Low pH. – The Firehole River. – Some Personal History. – Subject Index.

For ten years Prof. Brock and his associates carried out an extensive laboratory and field research program on the thermophilic microorganisms and life at high temperatures. Much of their research was done at Yellowstone National Park, but studies at all the major geothermal areas of the world were included in this effort. *Thermophilic Microorganisms and Life at High Temperatures* presents the results of this work; but it also reviews much of the other literature on the structure, function, ecology and practical significance of the thermophilic microorganisms.
This book will be of keen interest to all biologists, but especially to microbiologists. Biochemists, ecologists, environmental scientists and geologists will also find this work of great importance. The information presented has practical significance in such diverse fields as geothermal power, thermal pollution, exobiology, industrial enzymology and even solar energy production. It also is relevant to such basic and fundamental fields as evolution, the origin of life, paleomicrobiology and biogeochemistry.

G. Gottschalk
Bacterial Metabolism

1979. 161 figures, 41 tables. XI, 281 pages
ISBN 3-540-90308-9

Contents:
Nutrition of Bacteria. – How Escherichia coli Synthesizes ATP during Aerobic Growth on Glucose. – Biosynthesis of E. coli Cells from Glucose. – Aerobic Growth of E. coli on Substrates Other than Glucose. – Metabolic Diversity of Aerobic Heterotrophs. – Catabolic Activities of Aerobic Heterotrophs. – Regulation of Bacterial Metabolism. – Bacterial Fermentations. – Chemolithotrophic and Phototrophic Metabolism. – Fixation of Molecular Nitrogen. – Further Reading. – Index of Organisms. – Subject Index.

This text provides a survey of bacterial metabolism and describes its various facets in terms useful to students and researchers. Those reactions occuring only in microorganisms, or those ofe particular importance for them, are discussed with special emphasis. The energy metabolism of the various groups of bacteria, of aerobic and anaerobic heterotrophs of chemolithotrophs is therefore described in detail. The various pathways used by microorganisms for the degradation of numerous organic compounds are outlined as well and the fixation of molecular nitrogen is discussed. The reader will also find chapters dealing with the biosynthesis of cellular constituents and with the regulation of bacterial metabolism.

Springer-Verlag
Berlin
Heidelberg
New York